NEUROSCIENCE
INTELLIGENCE
UNIT

NEURAL TRANSPLANTATION IN CEREBELLAR ATAXIA

Lazaros C. Triarhou, M.D., Ph.D.

Indiana University School of Medicine
Indianapolis, Indiana, U.S.A.

Springer-Verlag
Berlin Heidelberg GmbH

R.G. LANDES COMPANY
AUSTIN

NEUROSCIENCE INTELLIGENCE UNIT

NEURAL TRANSPLANTATION IN CEREBELLAR ATAXIA

R.G. LANDES COMPANY
Austin, Texas, U.S.A.

International Copyright © 1997 Springer-Verlag Berlin Heidelberg
Originally published by Springer-Verlag Berlin, Heidelberg, Germany in 1997
Softcover reprint of the hardcover 1st edition 1997

 Springer

International ISBN 978-3-662-22215-7

While the authors, editors and publisher believe that drug selection and dosage and the specifications and usage of equipment and devices, as set forth in this book, are in accord with current recommendations and practice at the time of publication, they make no warranty, expressed or implied, with respect to material described in this book. In view of the ongoing research, equipment development, changes in governmental regulations and the rapid accumulation of information relating to the biomedical sciences, the reader is urged to carefully review and evaluate the information provided herein.

Library of Congress Cataloging-in-Publication Data

Triarhou, Lazaros Constantinos, 1957–
    Neural transplantation in cerebellar ataxia / Lazaros C. Triarhou.
        p. cm. -- (Neuroscience intelligence unit)
    Includes bibliographical references and index.
    ISBN 978-3-662-22215-7   ISBN 978-3-662-22213-3 (eBook)
    DOI 10.1007/978-3-662-22213-3

    1. Intracerebral transplantation. 2. Cerebellar ataxia. 3. Fetal
brain--Transplantation. 4. Cerebellar ataxia--Animal models.
    I. Title. II. Series.
        [DNLM: 1. Cerebellar Ataxia -- therapy. 2. Brain Tissue Transplantation -- methods.
    3. Cerebellum -- transplantation. 4. Mice, Mutant Strains. WL 320 T821n 1996]
RL594.12.T75 1996
617.4'810592—dc20
DNLM/DLC
for Library of Congress   95-23860
                                                                                            CIP

Publisher's Note

R.G. Landes Company publishes six book series: *Medical Intelligence Unit, Molecular Biology Intelligence Unit, Neuroscience Intelligence Unit, Tissue Engineering Intelligence Unit, Biotechnology Intelligence Unit and Environmental Intelligence Unit.* The authors of our books are acknowledged leaders in their fields and the topics are unique. Almost without exception, no other similar books exist on these topics.

Our goal is to publish books in important and rapidly changing areas of bioscience and environment for sophisticated researchers and clinicians. To achieve this goal, we have accelerated our publishing program to conform to the fast pace in which information grows in bioscience. Most of our books are published within 90 to 120 days of receipt of the manuscript. We would like to thank our readers for their continuing interest and welcome any comments or suggestions they may have for future books.

Shyamali Ghosh
Publications Director
R.G. Landes Company

CONTENTS

1. Introduction 1

2. Cerebellar Structure and Development 5
   Cells of the Cerebellum 5
   Afferent Connections 8
   Efferent Connections 8
   Ontogeny of Cerebellar Relationships 8
   The Inferior Olivary Complex 13
   Cerebellar Amino Acid Transmitters and Receptors 14
   Trophic Factor Systems Involved in Cerebellar Ontogeny 16

3. The Human Cerebellar Ataxias 31
   Introduction 31
   Predominantly Spinal Degenerations 33
   Cerebellar—Brainstem Degenerations 34
   Predominantly Cerebellar Degenerations 35

4. Cerebellar Mutants in the Laboratory Mouse 45
   Introduction 45
   Lurcher *(Lc)* 45
   Nervous *(nr)* 48
   Purkinje Cell Degeneration *(pcd)* 50
   Reeler *(rl)* 55
   Staggerer *(sg)* 58
   Weaver *(wv)* 60
   Other Mutations 64

5. Neurological Mutant Mice as Genetic Models
   for Neuronal Transplantation 81
   Introduction 81
   Visual System Models 83
   Neuroendocrine System 86
   Hippocampal Grafts 86
   Mesotelencephalic Dopamine Projection System 87
   Cerebellar Models 89

6. Basic Studies on Cerebellar Tissue Transplantation 95
   Introduction 95
   Survival of Cerebellar Primordia, Histotypic Differentiation
      and Synaptogenesis of Cerebellar Grafts 96
   Blood-Brain Barrier of Cerebellar Grafts 100
   Mitotic Activity 100
   Histochemical Phenotypy of Transplanted Purkinje Cells 101

Physiological Activity of Transplanted Purkinje Cells 102
Migratory Phenomena 103
Glial Issues 103
Graft-Host Interactions 105

7. **Structural Integration of Cerebellar Grafts
   in Ataxic Mouse Mutants** 113
   Introduction 113
   Transplantation Studies in Staggerer Mutant Mice 113
   Cerebellar Transplantation in Weaver Mutant Mice 115
   Cerebellar Transplants in *pcd* Mutant Mice 116
   Cerebellar Transplants in Nervous Mutant Mice 122
   Cerebellar Transplants in Lurcher Mutant Mice 123

8. **Cerebellar Grafting and the Recovery of Function** 131
   Introduction 131
   Spontaneous Movement and Stance 131
   Equilibrium 132
   Motor Coordination and Fatigue Resistance 133
   Open-Field Activity 133
   Pathophysiological Considerations 138
   A Protocol for Further Studies to Assess the Extent and Limits
      of Functional Recovery 140
   The Cerebellum and Higher Brain Functions 144

9. **Clinical Potential** 149
   Clinical Neural Transplantation Trials in Human
      Neurodegenerative Conditions 149
   Open Issues and Future Directions of Cerebellar Transplantation 150
   Concluding Remarks 152

Index 159

PREFACE

Cerebellar ataxia is a failure in muscular coordination that results from a slow, progressive deterioration of neurons in the cerebellum. At present, there is no known cure. This monograph presents a comprehensive treatise on the use of fetal grafts of developing cerebellar neuroblasts to counteract the structural and functional deficits associated with cerebellar degeneration in experimental models of hereditary ataxia. It covers elements of basic anatomy and development of the cerebellum; an overview of the inherited cerebellar ataxias in humans and mutant mouse models; neural transplantation studies in genetic mouse models with neuronal degenerations; the properties of cerebellar tissue growth and differentiation after experimental intraocular or intracerebral transplantation; recent advances in knowledge on the structural integration of cerebellar grafts in mutant mice with heredodegenerative ataxia and the functional recovery of behavioral responses; and a discussion of open issues and future directions, including the potential use of the neural grafting technique in counteracting certain forms of human cerebellar ataxias.

Acknowledgments

The author wishes to thank all of his collaborators and co-authors on the original articles, for without their contribution this work would not have been possible. In addition, words of gratitude are extended to Drs. M. del Cerro, B. Azzarelli and G.M. Spinella for insightful discussions, as well as to the American Society for Neural Transplantation for providing a supportive and creative environment. The author's original research studies have been funded in part by a research award from the National Institute of Neurological Disorders and Stroke, U.S. Department of Health and Human Services.

Finally, thanks are expressed to Dr. R.G. Landes, Dr. R. Wise, M. Kelly and L. O'Neill of Landes Bioscience Publishers for their kind words of encouragement, understanding and care during the preparation of the manuscript.

CHAPTER 1

INTRODUCTION

Cerebellar ataxia is a failure in muscular coordination that results from a slow, progressive deterioration of neurons in the cerebellum. An estimated 150,000 people are affected by the hereditary ataxias and related disorders in the United States. At present, there is no known cure. In an experimental treatment aimed at reconstructing the damaged pathway through exogenous neuronal supplementation, genetically ataxic mice have been used for intracerebral grafting of genetically healthy cerebellar neuroblasts, and evidence has been obtained for graft-induced enhancement of behavioral responses after bilateral cerebellar grafts. Such results are encouraging and underscore the potential of the neural grafting technique in restoring cerebellar function. However, many of the pathological and biochemical mechanisms in the interaction between grafted tissue and the host brain need to be further elucidated in extensive experimental studies, and great caution must be used in contemplating the theoretical feasibility of a possible application in humans.

The first study in the modern era of neural transplantation was published in 1971 by Das and Altman.[1] Since then the field of neural transplantation, has virtually exploded and the publication of several books and meetings' proceedings attests to that effect.[2-14] Cerebellar transplantation in mouse models is of particular interest, as: *(i)* the cerebellar anatomy, physiology and development are well understood;[15-19] *(ii)* the properties of cerebellar growth in culture have been also studied extensively[20-21] and *(iii)* the laboratory mouse is one of the mammalian species whose genetics are known in great detail.[22-23]

In this monograph, a synopsis is presented of the recent history of cerebellar tissue transplantation over the past 25 years. The properties of growth and differentiation of cerebellar grafts placed intraocularly or intracranially are reviewed, as well as the interaction of heterotopic and orthotopic grafts with the host brain. Particular emphasis is placed on the use of ataxic mouse mutants as recipients of donor cerebellar tissue for the correction of their structural deficits and the functional recovery of behavioral responses.

The chapters that follow cover basic elements of cerebellar neurobiology, an overview of the spectrum of diseases that affect the cerebellum in humans, a description of the most widely used mutant mouse models, neural transplantation studies at large in genetic mouse models with neuronal degenerations, structural and functional aspects of the use of fetal grafts of developing cerebellar neuroblasts to counteract the deficits associated with cerebellar degeneration in experimental models of hereditary ataxias and a discussion of open issues and future directions.

REFERENCES

1. Das GD, Altman J. Transplanted precursors of nerve cells: Their fate in the cerebellums of young rats. Science 1971; 173:637-638.
2. Wallace RB, Das GD, eds. Neural Tissue Transplantation Research. New York-Berlin-Heidelberg-Tokyo: Springer-Verlag, 1983.
3. Sladek JR Jr, Gash DM, eds. Neural Transplants: Development and Function. New York: Raven Press, 1984.
4. Björklund A, Stenevi U, eds. Neural Grafting in the Mammalian CNS. Amsterdam: Elsevier; 1985.
5. Azmitia EC, Björklund A, eds. Cell and Tissue Transplantation into the Adult Brain. New York: The New York Academy of Sciences, 1987.
6. Gash DM, Sladek JR Jr, eds. Transplantation into the Mammalian CNS. Amsterdam-New York-Oxford: Elsevier, 1988.
7. Dunnett SB, Richards S-J, eds. Neural Transplantation: From Molecular Basis to Clinical Applications. Amsterdam-New York-Oxford: Elsevier, 1990.
8. Lindvall O, Björklund A, Widner H, eds. Intracerebral Transplantation in Movement Disorders: Experimental Basis and Clinical Experiences. Amsterdam-London-New York-Tokyo: Elsevier, 1991.
9. Dunnett SB, Björklund A, eds. Neural Transplantation: A Practical Approach. Oxford-New York-Tokyo: Oxford University Press, 1992.

10. Dunnett SB, Björklund A, eds. Functional Neural Transplantation. New York: Raven Press, 1994.

11. Sanberg PR, Wictorin K, Isacson O. Cell Transplantation for Huntington's Disease. Austin, TX: RG Landes Co., 1994.

12. Vrbová G, Clowry G, Nógrádi A et al. Transplantation of Neural Tissue into Spinal Cord. Austin, TX: RG Landes Co., 1994.

13. Kordower JH, Sanberg PR, eds. Neural Transplantation into the CNS. Cell Transplantation, Vol 4, No 1 (Special Issue). Tarrytown, NY: Pergamon, 1995.

14. Isacson O, Kordower JH, eds. Neural Transplantation. Cell Transplantation Vol 5, No 2 (Special Issue). Tarrytown, NY: Elsevier, 1996.

15. Eccles JC, Ito M, Szentágothai J. The Cerebellum as a Neuronal Machine. Berlin, Heidelberg: Springer-Verlag, 1967.

16. Llinás R, ed. Neurobiology of Cerebellar Evolution and Development. Chicago: AMA/ERF Institute for Biomedical Research, 1969.

17. Palay SL, Chan-Palay V. Cerebellar Cortex: Cytology and Organization. Berlin-Heidelberg: Springer-Verlag, 1974.

18. Llinás R, Sotelo C, eds. The Cerebellum Revisited. New York-Berlin-Heidelberg: Springer-Verlag, 1992.

19. Altman J, Bayer SA. Development of the Cerebellar System: Evolution, Structure, and Functions. Boca Raton, FL: CRC Press, 1997.

20. Allerand CD. Patterns of neuronal differentiation in developing cultures of neonatal mouse cerebellum: A living and silver impregnation study. J Comp Neurol 1971; 142:167-204.

21. Seil FJ. Cerebellum in tissue culture. Rev Neurosci 1979; 4:105-177.

22. Sidman RL, Green MC, Appel SH. Catalog of the Neurological Mutants of the Mouse. Cambridge, MA: Harvard University Press, 1965.

23. Lyon MF, Searle AG, eds. Genetic Variants and Strains of the Laboratory Mouse. 2nd ed. Oxford, Stuttgart: Oxford University Press, Gustav Fischer Verlag, 1989.

CHAPTER 2

CEREBELLAR STRUCTURE AND DEVELOPMENT

CELLS OF THE CEREBELLUM

The cerebellum is a brain structure primarily involved in motor coordination. The adult cerebellar circuit is a product of precisely timed mitotic, migratory and synaptogenetic events during development.[1-21] The cerebellar cortex is a trilaminar structure (Fig. 2.1) containing five classes of neurons.[22-27] Stellate and basket cells are located in the superficial molecular layer, beneath which lies the layer of Purkinje cells. The internal granule cell layer is the deepest layer of the cerebellar cortex and contains granule and Golgi cells. During development, a transient external germinal layer is found superficially to the molecular layer (Fig. 2.2); this is where granule cells are generated and then migrate inbound to settle in their final location in the internal granule cell layer. The adult mouse cerebellum has been estimated to contain about 200,000 Purkinje cells and around 2×10^7 granule cells.[28,29] Specific immunochemical markers are available for the identification of Purkinje cells.[30-32]

Deep within the cerebellar white matter lie the deep cerebellar nuclei (Fig. 2.1), which in the mouse comprise the nucleus medialis, nucleus interpositus and nucleus lateralis (Fig. 2.3). The deep cerebellar nuclei of the adult mouse contain an estimated 10,000 neurons in each cerebellar hemisphere.[33] In the human cerebellum there are four nuclei in each hemisphere, the globose, emboliform, fastigial and dentate nucleus.

Fig. 2.1. Low-power light micrograph of the cerebellar hemisphere from a normal mouse cut parasagittally (upper). *One micrometer thick Epon section* (middle), *stained with toluidine blue, showing from top to bottom the molecular layer, the Purkinje cells and the granule cell layer. Ten micrometer thick paraffin section* (lower), *stained with gallocyanin, showing the nucleus lateralis at higher power. Magnification x25* (upper), *x400* (middle), *x100* (lower). *Upper and lower micrographs reprinted with permission from: Triarhou LC, Norton J, Ghetti B. Exp Brain Res 1987; 66:577-588. Middle micrograph reprinted with permission from: Triarhou LC, Low WC, Ghetti B. Anat Embryol 1987; 176: 145-154. © 1987 Springer-Verlag.*

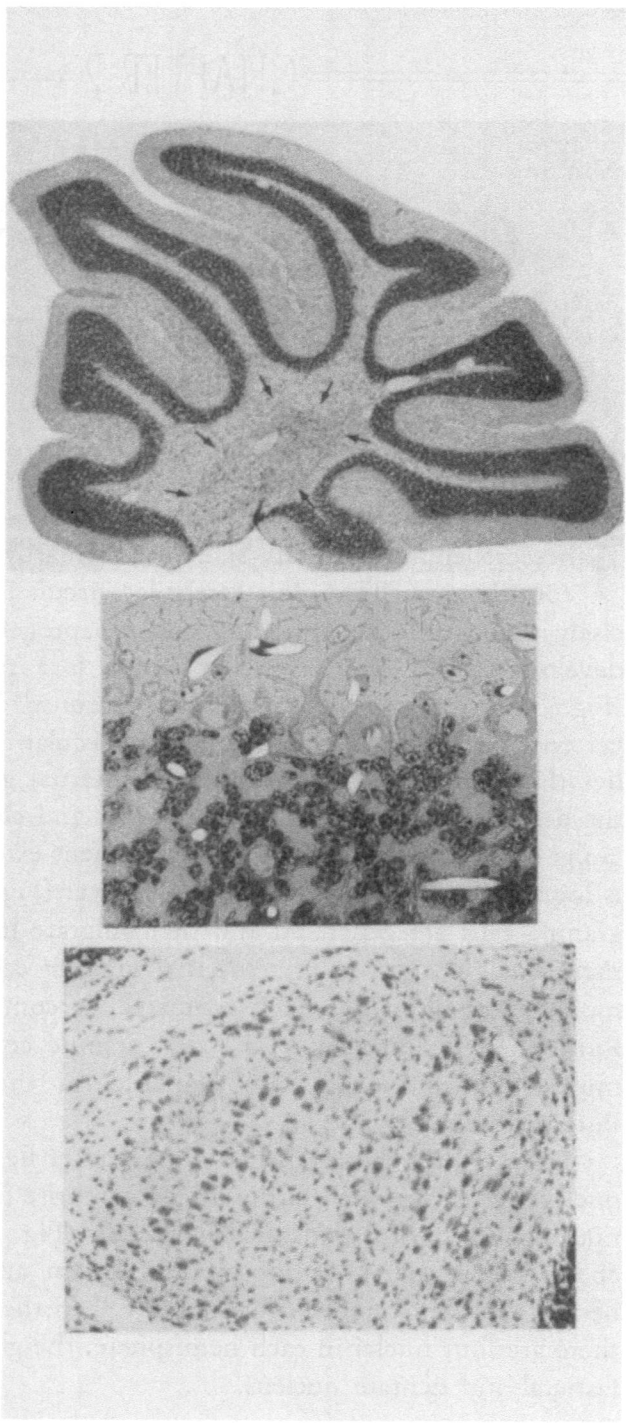

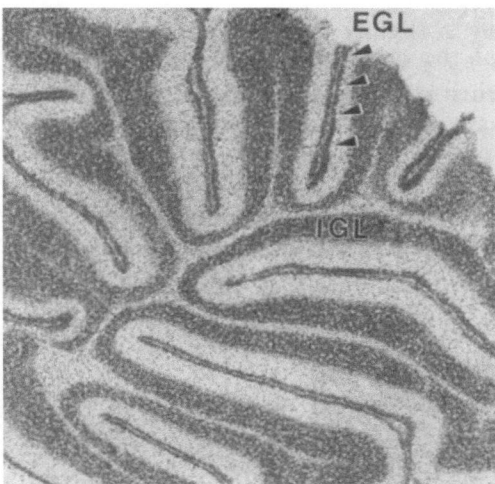

Fig. 2.2. Histological sagittal section of normal mouse cerebellum at 11 days of age to show the external germinal layer (EGL), where granule cells are produced before migrating inbound to form the internal granule cell layer (IGL). Magnification x70. Reprinted with permission from: Lee W-H, Wang G-M, Lo T et al. Mol Brain Res 1995; 30:259-268. © 1995 Elsevier Science B.V.

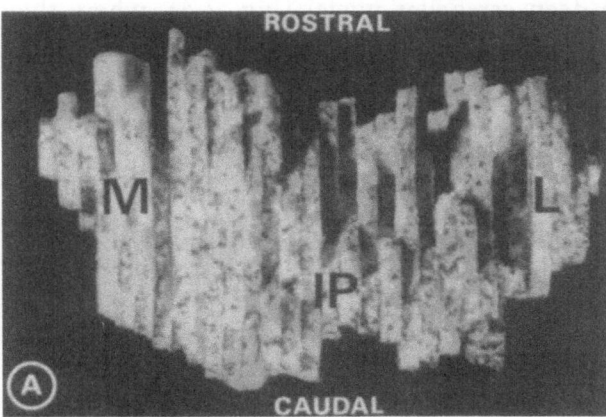

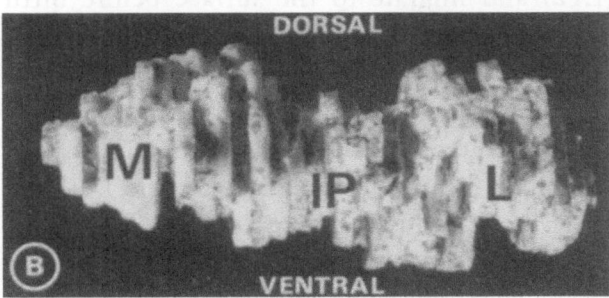

Fig. 2.3. Three-dimensional silicone models of the deep cerebellar nuclei in normal mouse. Views are of the inferior (A) and front (B) surfaces. Abbreviations: M, nucleus medialis; IP, nucleus interpositus; L, nucleus lateralis. Magnification x50. Reprinted with permission from: Triarhou LC, Norton J, Ghetti B. Exp Brain Res 1987; 66:577-588. ©1987 Springer-Verlag.

AFFERENT CONNECTIONS

The afferent innervation of the cerebellum consists of climbing fibers originating in the inferior olivary complex,[34-37] mossy fibers originating in pontine and spinal nuclei,[25,38] noradrenergic axons originating in the dorsal part of the nucleus locus coeruleus as well as in neurons of fields A5-A7 and the nucleus subcoeruleus,[39-43] and serotoninergic axons originating in the dorsal raphe nuclei of the pons and the medullary and pontine reticular formation.[44-47] All of these incoming afferents send collaterals both to the cerebellar cortex and to the deep cerebellar nuclei.

EFFERENT CONNECTIONS

The Purkinje cell is the only projection neuron of the cerebellar cortex. The remaining four types of nerve cells constitute local interneurons. Purkinje cell axons transmit signals from the cerebellar cortex to the deep cerebellar nuclei (Fig. 2.4), where they exert a powerful inhibition mediated by the neurotransmitter γ-aminobutyric acid (GABA).[48] In turn, axons of deep nuclei neurons transmit impulses to postcerebellar targets that include the ventrolateral nucleus of the thalamus, the red nucleus and the vestibular nuclei.[49] A small proportion of Purkinje cell axons project directly outside the cerebellum, to the dorsal part of the lateral vestibular nucleus of Deiters.

ONTOGENY OF CEREBELLAR RELATIONSHIPS

Purkinje cells are generated in the cerebellar primordium around embryonic day 12 (E12) and migrate to the surface before birth in the mouse.[1] Around postnatal day 3 (P3), Purkinje cells start to disperse in a monolayer and soon afterwards receive synaptic contacts from afferent axons. The advent of the interaction with migrating granule cells accelerates a profuse synaptogenesis with Purkinje cell dendrites which grow into the characteristic Purkinje cell dendritic trees by P12.[5,12] Purkinje cell maturation is thought to be a combination of genetic programming and an interaction with the developing cerebellar microenvironment. Golgi cells are generated toward the end of gestation, stellate and basket cells are produced during the first postnatal week and granule cells during the first two weeks of postnatal life.[1,2,12]

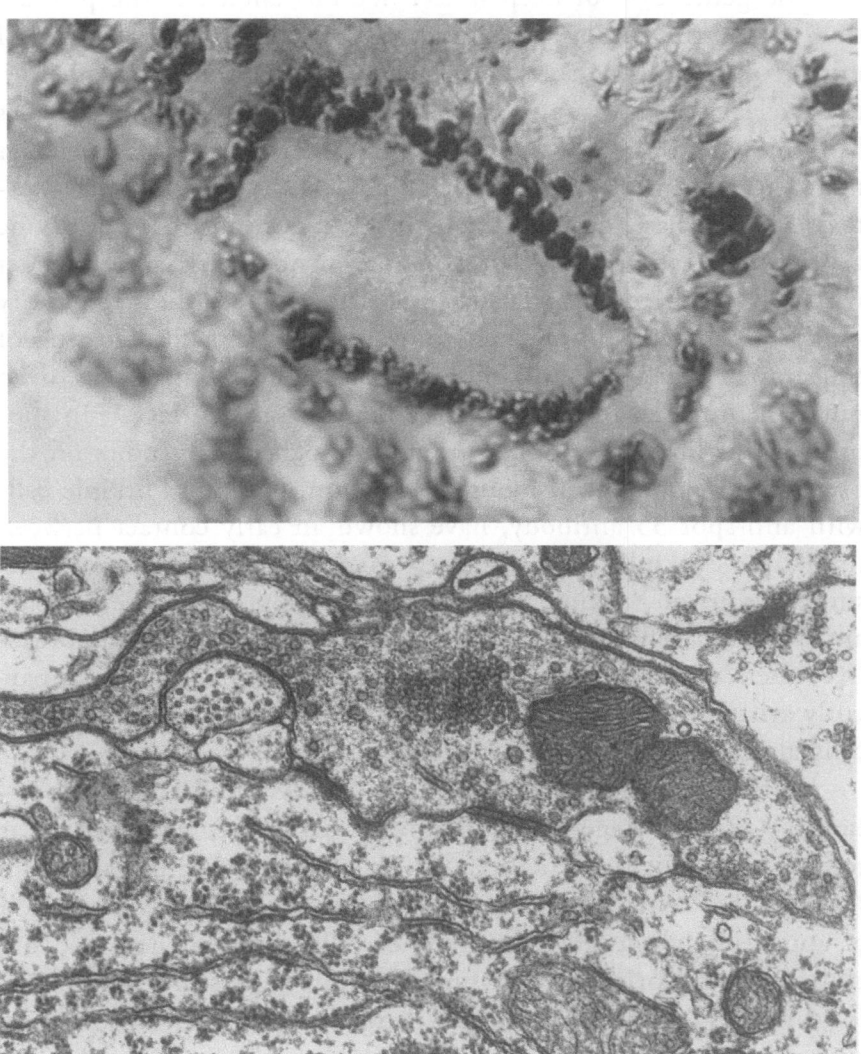

Fig. 2.4. Purkinje axon terminals, specifically immunolabeled with antibodies against 28 kDa Ca^{2+}-binding protein (CaBP), abutting the unlabeled soma of a deep nucleus neuron in mouse cerebellum (upper). Transmission electron micrograph (lower) of a Purkinje axon terminal forming axosomatic synaptic contact with a neuron of the deep cerebellar nucleus. Magnification x2500 (upper), x40000 (lower). Upper micrograph unpublished. Lower micrograph reprinted with permission from: Triarhou LC, Norton J, Ghetti B. Exp Brain Res 1987; 66:577-588. © 1987 Springer-Verlag.

The generation of deep nuclei neurons antedates the production of Purkinje cells by about a day.[12,15] After the cessation of mitotic divisions, the somata of deep nuclei neurons descend from the nuclear transitory zone to the depth of the cerebellum, whereas Purkinje cells follow a course of migration from the ventricular neuroepithelium of the cerebellar anlage to the cerebellar cortex.[1,15,16,50] It has been suggested that the portion of the Purkinje cell which is to become the axon may maintain synaptic contact with cells of the deep cerebellar nuclei from the earliest migratory stages of cerebellar histogenesis onwards[1] and that the crossing of trajectories may allow Purkinje cells to establish contact with deep nuclei neurons en route of their perikarya to the surface.[12] In support of this line of reasoning, immunocytochemical studies on the prenatal development of mouse cerebellum, marking Purkinje cells with anti-spot 35 antibody, have shown an early contact between immature Purkinje cells and deep nuclei cells.[51] After E16, presumed axons of Purkinje cells reach the region of the deep nuclei, and Purkinje cells move through it at E17. Spot 35-immunoreactive fibers enter the region of the deep cerebellar nuclei, apparently arising from both migratory Purkinje cells and from Purkinje cells already settled in the cerebellar cortex.[51]

By E15, the deep nuclear complex has acquired a medial position.[13,16] At this stage of embryonic development, a lamination is recognizable (Figs. 2.5 and 2.6). The following three layers are seen: (i) the "external germinal layer" dorsally, which is composed of approximately seven-eight layers of cells; some of the nuclei are ellipsoidal and oriented horizontally, while others are round; mitotic figures are present. (ii) A "marginal zone" lies beneath the external germinal layer and contains a few cellular components. (iii) Ventrally to the marginal zone is the "Purkinje plate", which contains six-seven layers of cells with round or oval nuclei. A thin, cell-free layer occupies the area immediately beneath the Purkinje plate, which is referred to as the "sub-Purkinje plate". Deeper in the cerebellar primordium, between the sub-Purkinje plate and the ventricular cavity, a thick area is rich in cells with nuclei of various shapes and dimensions. Mitotic figures are observed close to the ventricular surface. Blood vessels the caliber of 5-20 μm are

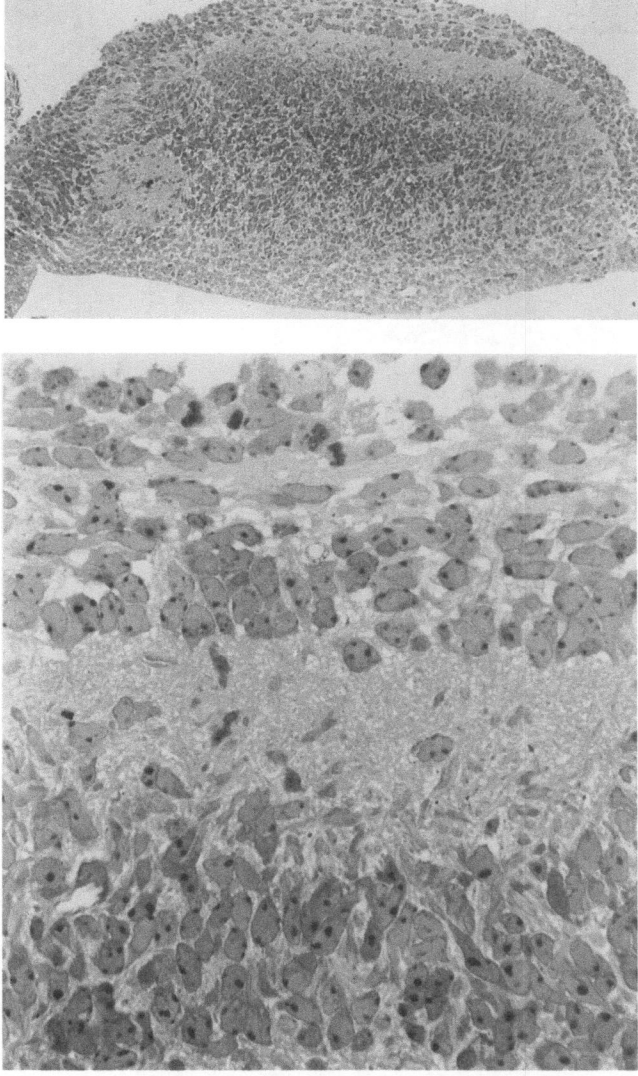

Fig. 2.5. Mouse cerebellar hemiprimordium (upper) at gestational day 15. High-power view of the primordial cerebellar cortex (lower) showing the external germinal layer on the top one-third of the field, the marginal zone in the middle and the Purkinje plate in the bottom one-third of the field; several mitotic figures are seen in the external germinal layer. One micrometer thick Epon sections stained with toluidine blue. Magnification ×100 (upper), ×630 (lower). Reprinted with permission from: Triarhou LC, Low WC, Ghetti B. Anat Embryol 1987; 176:145-154. © 1987 Springer-Verlag.

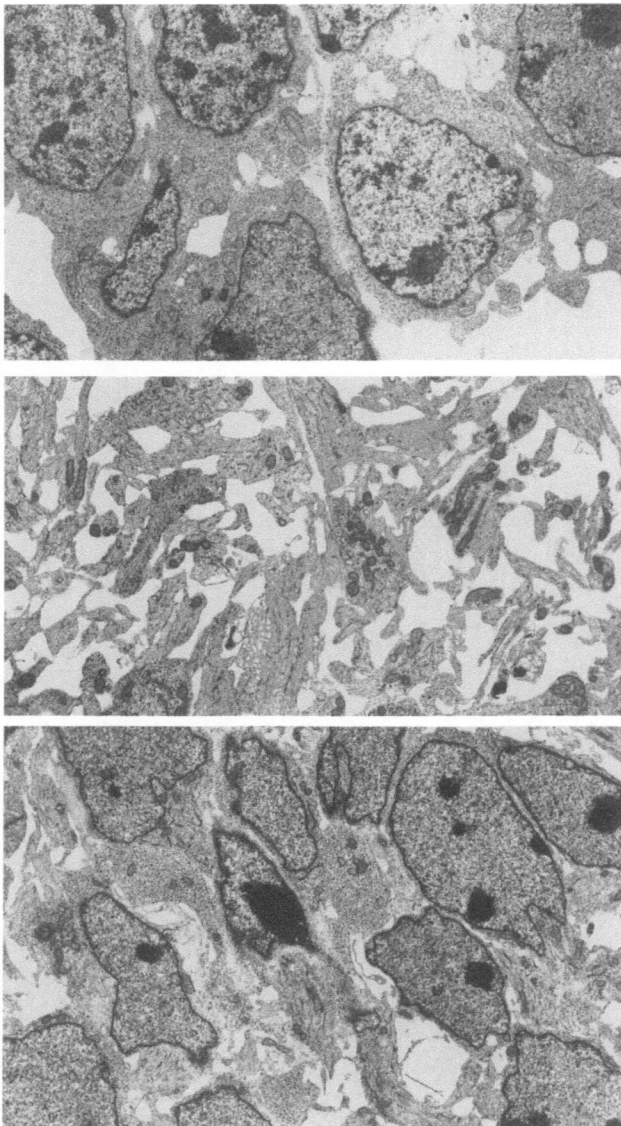

Fig. 2.6. Transmission electron micrographs showing in more detail the morphological appearance of cerebellar precursor cells in the external germinal layer (upper) and the Purkinje plate (lower) and of cellular processes in the marginal zone (middle) in the primordial cerebellar cortex of gestational day 15 mouse cerebellum. Uranyl acetate and lead citrate. Magnification ×4100 (upper), ×4400 (middle), ×3300 (lower). Original unpublished micrographs by the author.

distributed throughout the entire dorsoventral extent, but they are less frequent in the external germinal layer. In mice, Purkinje cells reach maturity by P15,[4,52] by which time cerebellar synaptogenesis is virtually complete.[5]

The stages of production of the remaining neuron classes of the cerebellum have been determined in detail in extensive [³H]thymidine-dating experiments in rats.[9,12,16,21,53,54] Golgi cells are generated between E19-P3, basket cells between P1-P11, stellate cells between P4-P15 and granule cells between P4-P19. Developmental neurogenetic timetables have been extrapolated for the human cerebellum, in which the deep cerebellar nuclei are thought to originate at 5-7 weeks of gestation, Purkinje cells at 6-7 weeks, Golgi cells at 10-24 weeks, basket cells at 24-32 weeks, stellate cells at 28-36 weeks and granule cells at 24-40 weeks.[55]

THE INFERIOR OLIVARY COMPLEX

The inferior olivary complex (Fig. 2.7) is the major source of olivocerebellar climbing fibers,[34,36,37] which establish synaptic contact with Purkinje dendritic thorns.[23-25,35] Synaptic connections also exist between climbing fiber collcaterals and cerebellocortical interneurons[25] and with neurons of the deep cerebellar nuclei.[56] The organization of olivocerebellar maps in the rodent cerebellum is discontinuous with sharp boundaries; Purkinje cells also constitute a heterogeneous population arranged in parasagittal compartments.[57]

The inferior olivary complex of the adult mouse contains about 25,000 neurons on both sides[58] and the putative neurotransmitter of climbing fibers is aspartate.[59,60] In the mouse, inferior olivary neurons are generated on E9-E11[61] and in the rat on E13-E14.[62] Migration of inferior olivary cells to their final location is complete by E16 in the mouse[63] and by E19 in the rat.[62] The postnatal development of the inferior olivary complex has been studied extensively in the rat,[64] while the corresponding information on mice can be extracted from the control animals in studies with mutant strains.[65-69] One important aspect of olivocerebellar development is a transient stage of multiple innervation of Purkinje cells by the climbing fibers, which gives way to a monoinnervation on P15 in the rat.[70-72]

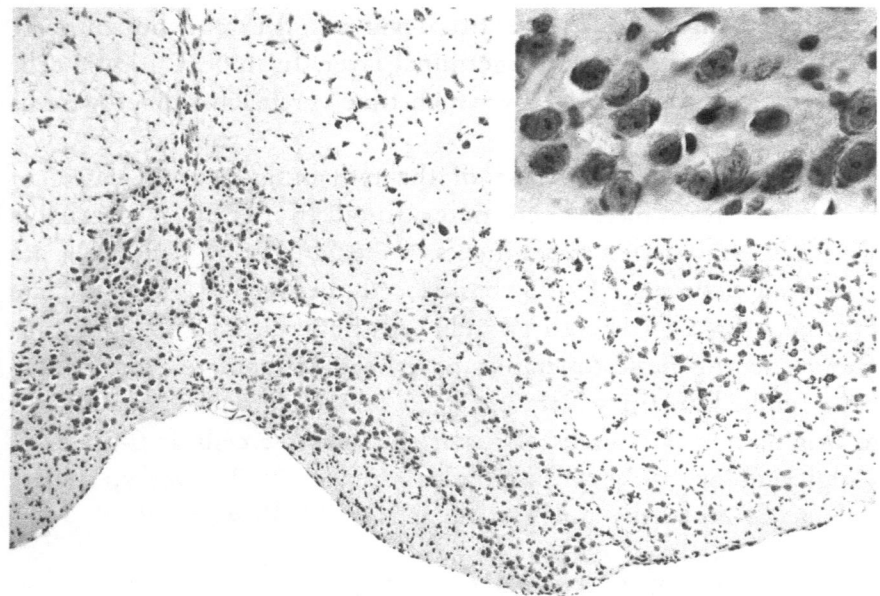

Fig. 2.7. Low-power light micrograph of paraffin sections of the mouse medulla oblongata cut coronally at the level of the inferior olivary complex. Inset shows high-power detail of individual neurons from principal olive. Magnification ×63, ×630 (inset). Reprinted with permission from: Ghetti B, Norton J, Triarhou LC: J Comp Neurol 1987; 260:409-422 © 1987 Alan R. Liss, Inc.

CEREBELLAR AMINO ACID TRANSMITTERS AND RECEPTORS

All types of cerebellar cortical neurons utilize GABA as their neurotransmitter, except granule cells, which use glutamate.[48,73]

Five different classes of subunits have been identified for the $GABA_A$ receptor, but which isoforms are expressed in CNS and the functional consequences of various subunit combinations are not totally clear.[74] Some progress has been made by expressing subunit mRNAs in transient expression systems, e.g., benzodiazepine modulation of the response to GABA requires coexpression of the α, β and γ subunits.[75] Further complexity is introduced by alternative splicing: the γ_2 subunit of the $GABA_A$ receptor exists in two forms, one of which lacks a critical eight residues that confer ethanol sensitivity on the receptor.[76] One parameter that has been used to evaluate a benzodiazepine-binding site in the α_1 subunit of the $GABA_A$ receptor complex in normal and *pcd* mutant mice is the binding of [³H]flunitrazepam to cerebellar sections.[77,78]

Glutamate is the major excitatory neurotransmitter in the mammalian CNS. In cerebellum, glutamate is the transmitter used by the granule cell parallel fibers.[73,79] The excitatory action of glutamate is mediated by three subtypes of "ionotropic" receptors (i.e. coupled to cation channels), named according to the agonists that preferentially excite them:[80-82] the NMDA receptor (*N*-methyl-D-aspartate, measured with ligands such as [³H]glutamate, [³H]glycine, and [³H]dizocilpine maleate or [³H]MK-801), and the "non-NMDA" receptors for kainic acid (measured by [³H]kainate) and AMPA (quisqualate-sensitive, measured by [³H](*RS*)-α-amino-3-hydroxy-5-methylisoxazole-4-propionate). AMPA has both high and low affinity binding sites. The non-NMDA antagonist 6-cyano-7-nitro-quinoxaline-2,3-dione (CNQX) binds with a single affinity to both the high and low affinity AMPA binding sites.[83] The fourth type of quisqualate-preferring "metabotropic" receptor coupled to a G protein is linked to inositol phospholipid metabolism.[84] L-[³H]glutamic acid and selective incubation conditions measure both the ionotropic and metabotropic quisqualate-sensitive receptors. A fifth receptor class, the L-AP4 receptor, has not yet been characterized with radioligand binding techniques.[85]

Cloning efforts have identified the individual components of these receptors.[86-90] The AMPA receptor can be recognized by antibodies made to its subunits.[91-94]

Non-NMDA glutamate receptors, including both the AMPA and kainic acid ionotropic subtypes, are localized on Purkinje cells, which receive an excitatory glutamate innervation from parallel fibers. [³H]CNQX binding sites are densely localized in the molecular layer of normal mouse cerebellum in association with Purkinje cell dendritic trees.[95] Analysis of equilibrium binding data in the molecular layer supports a single class of sites, and the K_i and B_{max} values for the inhibition of [³H]CNQX binding by various standard compounds in competition experiments indicate a higher affinity for kainic acid than AMPA and a biphasic displacement curve.[96] In wild-type mouse cerebellum, the B_{max} value of the granule cell layer is more than double that of the molecular layer.[97]

TROPHIC FACTOR SYSTEMS INVOLVED
IN CEREBELLAR ONTOGENY

Several trophic factor systems are known to be involved in various stages of cerebellar ontogeny and maintenance, including Purkinje cell morphogenesis. In particular, insulin-like growth factor-I (IGF-I),[98-105] basic fibroblast growth factor (bFGF),[106-109] the neurotrophins nerve growth factor (NGF),[110-113] brain-derived neurotrophic factor (BDNF) and neurotrophin-3 (NT3),[114-116] as well as several other factors such as epidermal growth factor (EGF),[106] homodimer of platelet-derived growth factor β-chain (PDGF),[117] ciliary neurotrophic factor (CNF),[118] growth hormone (GH) receptor binding protein,[119] pleiotrophin and midkine of the novel heparin-binding growth factor family,[120] have been shown to be expressed in developing cerebellum or to participate in some aspect of cerebellar neuron maturation and maintenance.

IGF-I is a pleiotrophic factor essential for cell proliferation, differentiation and survival in developing brain.[121] The most convincing evidence for IGF-I action has been found in cerebellum, where IGF-I has a dose-dependent, growth-promoting activity in cultured cerebellar cells, including Purkinje cells.[100,122] In vivo, IGF-I can be taken up, orthogradely transported and released by inferior olivary neurons into the cerebellar cortex.[123] During development, IGF-I is expressed exclusively by Purkinje cells. IGF-I mRNA levels increase after birth, reach a maximum around P11-P14 and decrease with maturity.[124] The expression of IGF-I, IGFR-I, IGFBP2 and IGFBP5 is synchronized with Purkinje cell maturation. Each IGF system gene is exclusively expressed by a specific cell population, except for IGFR-I mRNA, which is present in all cerebellar cells. Although IGF-I gene expression is coordinated with the growth activity of Purkinje cells, the functional role of IGF-I in Purkinje cell growth is only partially understood. The colocalization of IGF-I and IGFR-I mRNA in Purkinje cells[125] in both normal development and in cerebellar transplants suggests that IGF-I may act as an autocrine trophic factor required for Purkinje cell dendritic growth, synaptogenesis and myelination. IGF-I may also act as a paracrine growth factor on other cerebellar cells where the IGFR-I gene is expressed. There, IGF-I, released from Purkinje cells, might bind to IGFR-I to exert nonspecific trophic functions, such as regulating cellular metabolism upon demand.[102] In addition, IGF-I may act synergistically with other neurotrophic factors, such as NGF,[126] bFGF[127,128] and EGF.[129] On the other hand, IGF-I may act as a survival factor for cerebellar precursor cells, as shown by a dose-dependent increase of cultured cerebellar neurons with enhanced neurotransmitter synthesis,[100] and by an increase in newly formed oligodendrocytes and their precursors in culture.[130]

Basic FGF is a member of a family of structurally and functionally related pluripotent polypeptides with mitogenic and trophic action on cells of mesodermal and neuroectodermal origin.[131] The cDNA clone encoding rat bFGF has been cloned and sequenced and corresponds to a molecule consisting of 154 amino acid residues.[132] The effects of bFGF are mediated through binding to a high-affinity surface receptor (FGF-R) of 130 kDa, the complementary DNA of which has been cloned and sequenced.[133,134] The FGF-R contains three extracellular immunoglobulin-like domains, an unusual acidic region, and an intracellular tyrosine kinase domain, arranged in a pattern different from any growth factor receptor described.[133] The FGF-R sequence lies between the nucleotide sequences of two highly conserved amino acid motifs from the catalytic domain of protein-tyrosine kinases.[134] In the cerebellum, bFGF peptide is present in the somata, dendrites and axons of Purkinje cells from P7 on.[108,109] High-affinity [^{125}I]bFGF binding sites corresponding to a heparitinase-resistant receptor type are localized on neural process layers.[107] Further, bFGF has the ability of protecting cerebellar neurons in primary culture from excitatory amino acid neurotoxicity via NMDA and ionotropic non-NMDA receptors, possibly by modulating the extracellular formation of Ca^{2+}-dependent cGMP.[135]

NGF, the first neurotrophic factor to be identified, has an important role in the developing and mature nervous system and acts by binding to specific cell surface receptors;[136] therefore, definition of its receptor is fundamental to understanding its biological functions. The *trk* family of tyrosine kinase receptors have been identified as mediators in signal transduction by neurotrophins.[137,138] Low-affinity NGF-R, on the other hand, was shown to bind with similar (low) affinity to NGF, BDNF and NT3.[139] High-affinity of NGF seems to be dependent on the presence of both low-affinity NGF-R and *trk*A.[140] However, *trk*A alone was demonstrated as a high-affinity receptor for NGF.[137] Although low-affinity NGF-R alone is apparently unable to initiate signal transduction, modulatory roles in *trk*-mediated responses, as well as involvement in pathways not related to *trk*s have been proposed.[138] In cultured cerebellar neuroblasts, NGF enhances DNA synthesis in a dose-dependent manner, suggesting that the cerebellum may be a very good autocrine model of NGF action.[113] Furthermore, NGF in combination with glutamate, aspartate, high K^+ or veratridine promotes an increase in Purkinje cell survival, Purkinje cell size and neurite elaboration in dissociated cell cultures.[111] On the other hand, Purkinje cells express both low- and high-affinity NGF receptors, thus supporting a regulatory role of NGF in the morphogenesis of Purkinje cells.[110,112,114,141,142]

BDNF is a dimeric protein of 13.5 kDa subunits that is produced mainly in the CNS.[143] It has been shown to stimulate autophosphorylation of and transduce signals through the *trk*B member of the tyrosine kinase receptor family: the *trk* proto-oncogene is transcribed only in neural crest-derived components of the nervous system;[144] *trk*B, a gene structurally related to *trk*, is expressed in embryonic and adult nervous systems and has been identified as the receptor for BDNF.[145] The spatio-temporal pattern of expression of the BDNF gene in rat cerebellum suggests a maintenance role during postnatal development.[114] Further, BDNF treatment of primary cerebellar cultures induces RNA production of *Pax-2*, *Pax-3* and *Pax-6*, genes that encode sequence-specific DNA binding transcription factors expressed in embryonic development of the nervous system.[116] Finally, gene expression of BDNF has been detected by in situ hybridization histochemistry in the areas of grafted Purkinje and granule cells in the adult rat cerebellum.[115]

Glial cell line-derived neurotrophic factor (GDNF), a member of the type β transforming growth factor superfamily, has been shown to promote the survival and morphological differentiation of Purkinje cells with high potency in dissociated cultures of rat cerebellum;[146] moreover, GDNF RNA transcripts are expressed by granule cells and deep nuclei neurons from E17 onward, being maintained through adulthood.[146]

REFERENCES

1. Miale IL, Sidman RL. An autoradiographic analysis of histogenesis in the mouse cerebellum. Exp Neurol 1961; 4:277-296.
2. Fujita S. Quantitative analysis of cell proliferation and differentiation in the cortex of the postnatal mouse cerebellum. J Cell Biol 1967; 32:277-288.
3. del Cerro M, Snider RS, Oster ML. Evolution of the extracellular space in immature nervous tissue. Experientia (Basel) 1968; 24:929-930.
4. Meller K, Glees P. The development of mouse cerebellum: A Golgi and electron microscopical study. In: Llinás R, ed. Neurobiology of Cerebellar Evolution and Development. Chicago: AMA/ERF Institute for Biomedical Research, 1969:783-801.
5. Larramendi LMH. Analysis of synaptogenesis in the cerebellum of the mouse. In: Llinás R, ed. Neurobiology of Cerebellar Evolution and Development. Chicago: AMA/ERF Institute for Biomedical Research, 1969:803-843.
6. del Cerro M, Snider RS. Axo-somatic and axo-dendritic synapses in the cerebellum of the newborn rat. Brain Res 1972; 43:561-586.
7. del Cerro M, Swarz JR. Prenatal development of Bergmann glial fibers in rodent cerebellum. J Neurocytol 1976; 5:669-676.
8. West MJ, del Cerro M. Early formation of synapses in the molecular layer of the fetal rat cerebellum. J Comp Neurol 1976; 165:137-160.
9. Altman J, Bayer SA. Prenatal development of the cerebellar system in the rat. I. Cytogenesis and histogenesis of the deep nuclei and the cortex of the cerebellum. J Comp Neurol 1978; 179:23-48.
10. Altman J, Bayer SA. Prenatal development of the cerebellar system in the rat. II. Cytogenesis and histogenesis of the inferior olive, pontine gray, and the precerebellar reticular nuclei. J Comp Neurol 1978; 179:49-76.
11. Sotelo C. Mutant mice and the formation of cerebellar circuitry. Trends Neurosci 1980; 3:33-36.
12. Altman J. Morphological development of the rat cerebellum and some of its mechanisms. Exp Brain Res [Suppl] 1982; 6:8-49.
13. Goffinet AM. The embryonic development of the cerebellum in normal and reeler mutant mice. Anat Embryol (Berl) 1983; 168:73-86.
14. Altman J, Bayer SA. Embryonic development of the rat cerebellum. I. Delineation of the cerebellar primordium and early cell movements. J Comp Neurol 1985; 231:1-26.
15. Altman J, Bayer SA. Embryonic development of the rat cerebellum. II. Translocation and regional distribution of the deep neurons. J Comp Neurol 1985; 231:27-41.

16. Altman J, Bayer SA. Embryonic development of the rat cerebellum. III. Regional differences in the time of origin, migration, and settling of Purkinje cells. J Comp Neurol 1985; 231:42-65.

17. Bayer SA. The development of the central nervous system. In: Wiggins RC, McCandless DW, Enna SJ, eds. Developmental Neurochemistry. Austin, TX: University of Texas Press, 1985:8-46.

18. Sotelo C. Cerebellar synaptogenesis: Mutant mice—neuronal grafting. J Physiol (Paris) 1991; 85:134-144.

19. Altman J. The early stages of nervous system development: Neurogenesis and neuronal migration. In: Björklund A, Hökfelt T, Tohyama M, eds. Handbook of Chemical Neuroanatomy, Vol. 10: Ontogeny of Transmitters and Peptides in the CNS. Amsterdam: Elsevier Science Publishers BV, 1992:1-31.

20. Goldowitz D, Eisenman LM. Genetic mutations affecting murine cerebellar structure and function. In: Driscoll P, ed. Genetically Defined Animal Models of Neurobehavioral Dysfunctions. Boston-Basel-Berlin: Birkhäuser, 1992:66-88.

21. Altman J, Bayer SA. Development of the Cerebellar System: Evolution, Structure and Functions. Boca Raton, FL: CRC Press, 1997.

22. Ramón y Cajal S. Textura del Sistema Nervioso del Hombre y los Vertebrados, Vols. I-III. Madrid: Moya, 1897-1904.

23. Eccles JC, Ito M, Szentágothai J. The Cerebellum as a Neuronal Machine. Berlin-Heidelberg: Springer-Verlag, 1967.

24. Mugnaini E. The histology and cytology of the cerebellar cortex. In: Larsell O, Jansen J, eds. The Comparative Anatomy and Histology of the Cerebellum: The Human Cerebellum, Cerebellar Connections, and Cerebellar Cortex. Minneapolis: University of Minnesota Press, 1972:201-264.

25. Palay SL, Chan-Palay V. Cerebellar Cortex: Cytology and Organization. Berlin-Heidelberg: Springer-Verlag, 1974.

26. Ito M. The Cerebellum and Neural Control. New York: Raven Press, 1984.

27. Triarhou LC, Low WC, Ghetti B. Transplantation of cerebellar anlagen to hosts with genetic cerebellocortical atrophy. Anat Embryol (Berl) 1987; 176:145-154.

28. Caddy KWT, Biscoe TJ. Structural and quantitative studies in the normal C3H and Lurcher mutant mouse. Phil Trans Roy Soc Lond (Biol) 1979; 287:167-201.

29. Harvey RJ, Napper RM. Quantitative study of granule and Purkinje cells in the cerebellar cortex of the rat. J Comp Neurol 1988; 274:151-157.

30. Mugnaini E, Morgan JI. The neuropeptide cerebellin is a marker for two similar neuronal circuits in rat brain. Proc Natl Acad Sci USA 1987; 84:8692-8696.

31. Celio MR, Baier W, Schärer L et al. Monoclonal antibodies directed against the calcium binding protein Calbindin D-28k. Cell Calcium 1990; 11:599-602.

32. Tokunaga A, Ono K, Date I et al. A monoclonal antibody that labels Purkinje cells in the rat cerebellum. Brain Res Bulletin 1991; 27:669-674.

33. Triarhou LC, Norton J, Ghetti B. Anterograde transsynaptic degeneration in the deep cerebellar nuclei of Purkinje cell degeneration (*pcd*) mutant mice. Exp Brain Res 1987; 66:577-588.

34. Szentágothai J, Rajkovits K. Über den Ursprung der Kletterfasern des Kleinhirns. Z Anat Entwickl-Gesch 1959; 121:130-141.

35. Hámori J, Szentágothai J. Identification under the electron microscope of climbing fibres and their synaptic contacts. Exp Brain Res 1966; 1:65-81.

36. Desclin JC. Histological evidence supporting the inferior olive as the major source of cerebellar climbing fibers in the rat. Brain Res 1974; 77:365-384.

37. Courville J, Faraco-Cantin F. On the origin of the climbing fibers of the cerebellum. An experimental study in the cat with an autoradiographic tracing method. Neuroscience 1978; 3:797-809.

38. Ramón y Cajal S. Sur les fibres moussues et quelques points douteux de la texture de l' écorce cérébelleuse. Trab Lab Invest Biol Univ Madrid 1926; 24:215-251.

39. Olson L, Fuxe K. On the projections from the locus coeruleus noradrenaline neurons: the cerebellar innervation. Brain Res 1971; 28:165-171.

40. Hoffer BJ, Siggins GR, Oliver AP et al. Activation of the pathway from locus coeruleus to rat cerebellar Purkinje neurons: Pharmacological evidence of noradrenergic central inhibition. J Pharmac Exp Ther 1973; 184:553-569.

41. Tohyama M. Comparative anatomy of cerebellar catecholamine innervations from teleosts to mammals. J Hirnforsch 1976; 17:43-60.

42. Pasquier DA, Gold MA, Jacobowitz DM. Noradrenergic perikarya (A5-A7, subcoeruleus) projections to the rat cerebellum. Brain Res 1980; 196:270-275.

43. Triarhou LC, Ghetti B. Monoaminergic nerve terminals in the cerebellar cortex of Purkinje cell degeneration mutant mice: Fine structural integrity and modification of cellular environs following loss of Purkinje and granule cells. Neuroscience 1986; 18: 795-807.

44. Taber Pierce E, Hoddevik GH, Walberg F. The cerebellar projection from the raphe nuclei in the cat as studied with the method of retrograde transport of horseradish peroxidase. Anat Embryol (Berl) 1977; 152:73-87.

45. Takeuchi Y, Kimura H, Sano Y. Immunohistochemical demonstration of serotonin-containing nerve fibers in the cerebellum. Cell Tissue Res 1982; 226:1-12.
46. Bishop GA, Ho RH. The distribution and origin of serotonin immunoreactivity in the rat cerebellum. Brain Res 1985; 331:195-207.
47. Triarhou LC, Ghetti B. Serotonin-immunoreactivity in the cerebellum of two neurological mutant mice and the corresponding wild-type genetic stocks. J Chem Neuroanat 1991; 4:421-428.
48. Obata K. GABA in Purkinje cells and motoneurons. Experientia (Basel) 1969; 25:1285.
49. Thach WT, Kane SA, Mink JW et al. Cerebellar output: Multiple maps and modes of control in movement coordination. In: Llinás R, Sotelo C, eds. The Cerebellum Revisited. New York: Springer-Verlag, 1992:283-300.
50. Ramón y Cajal S. Studies on Vertebrate Neurogenesis (1929). Guth L, transl. Springfield, IL: CC Thomas, 1960:251-321.
51. Yuasa S, Kawamura K, Ono K et al. Development and migration of Purkinje cells in the mouse cerebellar primordium. Anat Embryol (Berl) 1991; 184:195-212.
52. Hendelman WJ, Rouf K. The development of the Purkinje cell of the mouse: A Golgi analysis. Anat Rec 1974; 178:372.
53. Bayer SA, Altman J. Neurogenesis and neuronal migration. In: Paxinos G, ed. The Rat Nervous System. 2nd ed. Orlando, FL: Academic Press, 1995:1041-1078.
54. Bayer SA, Altman J. Principles of neurogenesis, neuronal migration, and neural circuit formation. In: Paxinos G, ed. The Rat Nervous System. 2nd ed. Orlando, FL: Academic Press, 1995: 1079-1098.
55. Bayer SA, Altman J, Russo RJ et al. Timetables of neurogenesis in the human brain based on experimentally determined patterns in the rat. Neurotoxicology 1993; 14:83-144.
56. Van Der Want JJL, Wiklund L, Guegan M et al. Anterograde tracing of the rat olivocerebellar system with *Phaseolus vulgaris* leucoagglutinin (PHA-L): Demonstration of climbing fiber collateral innervation of the cerebellar nuclei. J Comp Neurol 1989; 288:1-18.
57. Sotelo C, Wassef M. Cerebellar development: Afferent organization and Purkinje cell heterogeneity. Phil Trans Roy Soc Lond (Biol) 1991; 331:307-313.
58. Triarhou LC, Ghetti B. Stabilisation of neurone number in the inferior olivary complex of aged 'Purkinje cell degeneration' mutant mice. Acta Neuropathol (Berl) 1991; 81:597-602.

59. Nadi NS, Kanter D, McBride WJ et al. Effects of 3-acetylpyridine on several putative neurotransmitter amino acids in the cerebellum and medulla of the rat. J Neurochem 1976; 28:661-662.

60. Wiklund L, Toggenburger G, Cuénod M. Aspartate: Possible neurotransmitter in cerebellar climbing fibers. Science 1982; 216:78-80.

61. Taber Pierce E. Time of origin of neurons in the brain stem of the mouse. Prog Brain Res 1973; 40:53-65.

62. Altman J, Bayer SA. Development of the precerebellar nuclei in the rat: II. The intramural olivary migratory stream and the neurogenetic organization of the inferior olive. J Comp Neurol 1987; 257:490-512.

63. Goffinet AM. The embryonic development of the inferior olivary complex in normal and reeler (*rl* ORL) mutant mice. J Comp Neurol 1983; 219:10-24.

64. Bourrat F, Sotelo C. Postnatal development of the inferior olivary complex in the rat. III. A morphometric analysis of volumetric growth and neuronal cell number. Brain Res 1984; 16:241-251.

65. Shojaeian H, Delhaye-Bouchaud N, Mariani J. Neuronal death and synapse elimination in the olivocerebellar system. II. Cell counts in the inferior olive of adult X-irradiated rats and *weaver* and *reeler* mutant mice. J Comp Neurol 1985; 232:309-318.

66. Shojaeian H, Delhaye-Bouchaud N, Mariani J. Decreased number of cells in the inferior olivary nucleus of the developing staggerer mouse. Dev Brain Res 1985; 21:141-146.

67. Blatt GJ, Eisenman LM. A qualitative and quantitative light microscopic study of the inferior olivary complex of normal, reeler, and weaver mutant mice. J Comp Neurol 1985; 232:117-128.

68. Blatt GJ, Eisenman LM. A qualitative and quantitative light microscopic study of the inferior olivary complex in the adult staggerer mutant mouse. J Neurogenet 1985; 2:51-66.

69. Ghetti B, Norton J, Triarhou LC. Nerve cell atrophy and loss in the inferior olivary complex of "Purkinje cell degeneration" mutant mice. J Comp Neurol 1987; 260:409-422.

70. Mariani J, Changeux J-P. Étude par enregistrements intracellulaires de l' innervation multiple des cellules de Purkinje par les fimbres grimpantes dans le cervelet du rat en développement. C R Acad Sci (Paris) 1980; 291:97-100.

71. Mariani J, Changeux J-P. Ontogenesis of olivocerebellar relationships. I. Studies by intracellular recordings of the multiple innervation of Purkinje cells by climbing fibers in the developing rat cerebellum. J Neurosci 1981; 1:696-702.

72. Mariani J, Changeux J-P. Ontogenesis of olivocerebellar relationships. II. Spontaneous activity of inferior olivary neurons and climbing fiber-mediated activity of cerebellar Purkinje cells in developing rats. J Neurosci 1981; 1:703-709.

73. Hudson DB, Valcana T, Bean G et al. Glutamic acid: a strong candidate as the neurotransmitter of the cerebellar granule cell. Neurochem Res 1976; 1:73-81.
74. Taylor CW, Lummis SCR. An introduction to receptors. In: Wharton J, Polak JM, eds. Receptor Autoradiography: Principles and Practice. Oxford: Oxford University Press, 1993:1-12.
75. Pritchett DB, Sontheimer H, Shivers BD et al. Importance of a novel GABA$_A$ receptor and subunit for benzodiazepine pharmacology. Nature (Lond) 1989; 338:582-585.
76. Wafford KA, Burnett DM, Leidenheimer NJ et al. Ethanol sensitivity of the GABA$_A$ receptor expressed in *Xenopus* oocytes requires 8 amino acids contained in γ2L subunit. Neuron 1991; 7:27-33.
77. Kahle G, Kaulen P, Bruning G et al. Autoradiographic analysis of benzodiazepine receptors in mutant mice with cerebellar defects. J Chem Neuroanat 1990; 3:261-270.
78. Zdilar D, Rotter A, Frostholm A. Expression of GABA$_A$/benzodiazepine receptor α_1-subunit mRNA and [^{3}H]flunitrazepam binding sites during postnatal development of the mouse cerebellum. Dev Brain Res 1991; 61:63-71.
79. Fagg GE, Foster AC. Amino acid neurotransmitters and their pathways in the mammalian central nervous system. Neuroscience 1983; 9:701-719.
80. Wamsley JK, Palacios JM. Amino acid and benzodiazepine receptors. In: Björklund A, Hökfelt T, Kuhar MJ, eds. Handbook of Chemical Neuroanatomy, Vol 3: Classical Transmitters and Transmitter Receptors in the CNS, Part II. Amsterdam: Elsevier Science Publishers BV, 1984:352-385.
81. Foster AC, Fagg GE. Acidic amino acid binding sites in mammalian neuronal membranes: their characteristics and relationship to synaptic receptors. Brain Res Rev 1984; 7:103-164.
82. Monaghan DT, Bridges RJ, Cotman CW. The excitatory amino acid receptors: Their classes, pharmacology, and distinct properties in the function of the central nervous system. Ann Rev Pharmacol Toxicol 1989; 29:365-402.
83. Nielsen EØ, Drejer J, Cha JJ et al. Autoradiographic characterization and localization of quisqualate binding sites in rat brain using the antagonist [^{3}H]CNQX: comparison with [^{3}H]AMPA binding sites. J Neurochem 1990; 54:686-695.
84. Sugiyama H, Ito I, Hirono C. A new type of glutamate receptor linked to inositol phospholipid metabolism. Nature (Lond) 1987; 325:531-533.
85. Monaghan DT. Autoradiographic analysis of excitatory amino-acid receptors. In: Wharton J, Polak JM, eds. Receptor Autoradiography:

Principles and Practice. Oxford: Oxford University Press, 1993:171-193.

86. Hollmann M, O'Shea-Greenfield A, Rogers SW et al. Cloning by functional expression of a member of the glutamate receptor family. Nature (Lond) 1989; 342:643-648.

87. Keinänen K, Wisden W, Sommer B et al. A family of AMPA-selective glutamate receptors. Science 1990; 249:556-560.

88. Gasic GP, Hollmann M. Molecular neurobiology of glutamate receptors. Ann Rev Physiol 1992; 54:507-536.

89. Sommer B, Seeburg PH. Glutamate receptor channels: novel properties and new clones. Trends Pharmacol Sci 1992; 13:291-296.

90. Ohishi H, Shigemoto R, Nakanishi S et al. Distribution of the mRNA for a metabotropic glutamate receptor (mGluR3) in the rat brain: An in situ hybridization study. J Comp Neurol 1993; 335:252-266.

91. Wenthold RJ, Hunter C, Wada K et al. Antibodies to a C-terminal peptide of the rat brain glutamate receptor subunit, GluR-A, recognize a subpopulation of AMPA binding sites but not kainate sites. FEBS Letters 1990; 276:147-150.

92. Wenthold RJ, Yokotani N, Doi K et al. Immunochemical characterization of the non-NMDA glutamate receptor using subunit-specific antibodies: Evidence for a hetero-oligomeric structure in rat brain. J Biol Chem 1992; 267:501-507.

93. Petralia RS, Wenthold RJ. Light and electron immunocytochemical localization of AMPA-selective glutamate receptors in the rat brain. J Comp Neurol 1992; 318:329-354.

94. Ryo Y, Miyawaki A, Furuichi T et al. Immunohistochemical localization of metabotropic and ionotropic glutamate receptors in the mouse brain. Ann NY Acad Sci 1993; 707:554-556.

95. Makowiec RL, Cha JJ, Penney JB et al. Cerebellar excitatory amino acid binding sites in normal, granuloprival, and Purkinje cell-deficient mice. Neuroscience 1991; 42:671-681.

96. Zavitsanou K, Mitsacos A, Kouvelas ED. Autoradiographic characterization of [^{3}H]6-cyano-7-nitroquinoxaline-2,3-dione binding sites in adult chick brain. Neuroscience 1994; 63:955-962.

97. Kouvelas ED, Mitsacos A, Angelatou F et al. Glutamate receptors in mammalian cerebellum: Alterations in human ataxic disorders and cerebellar mutant mice. In: Plaitakis A, ed. Cerebellar Degenerations: Clinical Neurobiology. Boston: Kluwer Academic Publishers, 1992:123-137.

98. Andersson IK, Edwall D, Norstedt G et al. Differing expression of insulin-like growth factor I in the developing and in the adult rat cerebellum. Acta Physiol Scand 1988; 132:167-173.

99. Torres-Aleman I, Pons S, Santos-Benito FF. Survival of Purkinje cells in cerebellar cultures is increased by insulin-like growth factor I. Eur J Neurosci 1992; 4:864-869.

100. Torres-Aleman I, Pons S, Arevalo MA. The insulin-like growth factor I system in the rat cerebellum: Developmental regulation and role in neuronal survival and differentiation. J Neurosci Res 1994; 39:117-126.

101. Bondy C, Lee W-H. Correlation between insulin-like growth factor (IGF)-binding protein 5 and IGF-I gene expression during brain development. J Neurosci 1993; 13:5092-5104.

102. Bondy CA, Lee W-H. Developmental and injury-induced patterns of IGF and IGF receptor gene expression in the brain: functional implications. Ann NY Acad Sci 1993; 692:33-43.

103. Lee W-H, Javedan S, Bondy CA. Coordinate expression of insulin-like growth factor system components by neurons and neuroglia during retinal and cerebellar development. J Neurosci 1992; 12:4737-4744.

104. Lee PDK, Conover CA, Powell DR. Regulation and function of insulin-like growth factor-binding protein-1. Proc Soc Exp Biol Med 1993; 204:4-29.

105. Lee W-H, Michels KM, Bondy CA. Localization of insulin-like growth factor binding protein-2 mRNA during postnatal brain development: correlation with insulin-like growth factors I and II. Neuroscience 1993; 53:251-265.

106. Abe K, Takayanagi M, Saito H. Basic fibroblast growth factor and epidermal growth factor promote survival of primary cultured cerebellar neurons from neonatal rats. Jap J Pharmacol 1991; 56:113-116.

107. Fayein NA, Courtois Y, Jeanny JC. Basic fibroblast growth factor high and low affinity binding sites in developing mouse brain, hippocampus and cerebellum. Biol Cell 1992; 76:1-13.

108. Matsuda S, Okumura N, Yoshimura H et al. Basic fibroblast growth factor-like immunoreactivity in Purkinje cells of the rat cerebellum. Neuroscience 1992; 50:99-106.

109. Matsuda S, Ii Y, Desaki J et al. Development of Purkinje cell bodies and processes with basic fibroblast growth factor-like immunoreactivity in the rat cerebellum. Neuroscience 1994; 59:651-662.

110. Cohen-Cory S, Dreyfus CF, Black IB. Expression of high- and low-affinity nerve growth factor receptors by Purkinje cells in the developing rat cerebellum. Exp Neurol 1989; 105:104-109.

111. Cohen-Cory S, Dreyfus CF, Black IB. NGF and excitatory neurotransmitters regulate survival and morphogenesis of cultured cerebellar Purkinje cells. J Neurosci 1991; 11:462-471.

112. Wanaka A, Johnson EM. Developmental study of nerve growth factor receptor mRNA in the postnatal rat cerebellum. Dev Brain Res 1990; 55:288-292.

113. Confort C, Charrasse S, Clos J. Nerve growth factor enhances DNA synthesis in cultured cerebellar neuroblasts. Neuroreport 1991; 2:566-568.

114. Rocamora N, García-Ladona FJ, Palacios JM et al. Differential expression of brain-derived neurotrophic factor, neurotrophin-3, and low-affinity nerve growth factor receptor during the postnatal development of the rat cerebellar system. Mol Brain Res 1993; 17:1-8.

115. Tsurushima H, Yuasa S, Kawamura K et al. Expression of tenascin and BDNF during the migration and differentiation of grafted Purkinje and granule cells in the adult rat cerebellum. Neurosci Res 1993; 18:109-120.

116. Kioussi C, Grüss P. Differential induction of *Pax* genes by NGF and BDNF in cerebellar primary cultures. J Cell Biol 1994; 125:417-425.

117. Smits A, Ballagi AE, Funa K. PDGF-BB exerts trophic activity on cultured GABA interneurons from the newborn rat cerebellum. Eur J Neurosci 1993; 5:986-994.

118. Larkfors L, Lindsay RM, Alderson RF. Ciliary neurotrophic factor enhances the survival of Purkinje cells in vitro. Eur J Neurosci 1994; 6:1015-1025.

119. Lincoln DT, El-Hifnawi E, Sinowatz F et al. Immunohistochemical localization of growth hormone receptor binding protein in the mammalian cerebellum. Ann Anat 1994; 176:419-427.

120. Matsumoto K, Wanaka A, Mori T et al. Localization of pleiotrophin and midkine in the postnatal developing cerebellum. Neurosci Lett 1994; 178:216-220.

121. Hepler JE, Lund PK. Molecular biology of the insulin-like growth factors: Relevance to nervous system function. Mol Neurobiol 1990; 4:93-127.

122. Torres-Aleman I, Pons S, García-Segura LM. Climbing fiber deafferentation reduces insulin-like growth factor I (IGF-I) content in cerebellum. Brain Res 1991; 564:348-351.

123. Nieto-Bona MP, García-Segura LM, Torres-Aleman I. Orthograde transport and release of insulin-like growth factor I from the inferior olive to the cerebellum. J Neurosci Res 1993; 36:520-527.

124. Bondy CA. Transient IGF-I gene expression during the maturation of functionally related central projection neurons. J Neurosci 1991; 11:3442-3455.

125. Bondy C, Werner H, Roberts CT et al. Cellular pattern of type-I insulin-like growth factor receptor gene expression during maturation of the rat brain: Comparison with insulin-like growth factors I and II. Neuroscience 1992; 46:909-923.

126. Recio-Pinto E, Lang FF, Ishii DN. Insulin and insulin-like growth factor II permit nerve growth factor binding and the neurite formation response in cultured human neuroblastoma cells. Proc Natl Acad Sci USA 1984; 81:2562-2566.

127. Pons S, Torres-Aleman I. Basic fibroblast growth factor modulates insulin-like growth factor-I, its receptor, and its binding proteins in hypothalamic cell cultures. Endocrinology 1992; 131:2271-2278.

128. Torres-Aleman I, Naftolin F, Robbins RJ. Trophic effects of basic fibroblast growth factor on fetal rat hypothalamic cells: interactions with insulin-like growth factor I. Dev Brain Res 1990; 52:253-257.

129. Chernausek SD. Insulin-like growth factor-I (IGF-I) production by astroglial cells: Regulation and importance for epidermal growth factor-induced cell replication. J Neurosci Res 1993; 34:189-197.

130. Barres BA, Hart KI, Coles HSR et al. Cell death and control of cell survival in the oligodendrocyte lineage. Cell 1992; 70:31-46.

131. Unsicker K, Grothe C, Lüdecke G et al. Fibroblast growth factors: Their roles in the central and peripheral nervous system. In: Loughlin SE, Fallon JH, eds. Neurotrophic Factors. San Diego: Academic Press, 1993:313-338.

132. Shimasaki S, Emoto N, Koba A et al. Complementary DNA cloning and sequencing of rat ovarian basic fibroblast growth factor and tissue distribution study of its mRNA. Biochem Biophys Res Commun 1988; 157:256-263.

133. Lee PL, Johnson DE, Cousens LS et al. Purification and complementary DNA cloning of a receptor for basic fibroblast growth factor. Science 1989; 245:57-60.

134. Reid HH, Wilks AF, Bernard O. Two forms of the basic fibroblast growth factor receptor-like mRNA are expressed in the developing mouse brain. Proc Natl Acad Sci USA 1990; 87:1596-1600.

135. Fernandez-Sanchez MT, Novelli A. Basic fibroblast growth factor protects cerebellar neurons in primary culture from NMDA and non-NMDA receptor mediated neurotoxicity. FEBS Lett 1993; 335:124-131.

136. Longo FM, Holtzman DM, Grimes ML et al. Nerve growth factor: Actions in the peripheral and central nervous systems. In: Loughlin SE, Fallon JH, eds. Neurotrophic Factors. San Diego: Academic Press, 1993:209-256.

137. Klein R, Jing S, Nanduri V et al. The *trk* proto-oncogene encodes a receptor for nerve growth factor. Cell 1991; 65:189-197.

138. Glass DJ, Nye SH, Hatzopoulos P et al. *Trk*B mediates BDNF/NT3-dependent survival and proliferation in fibroblasts lacking the low affinity NGF receptor. Cell 1991; 66:405-413.

139. Rodriguez-Tébar A, Dechant G, Barde Y-A. Binding of brain-derived neurotrophic factor to the nerve growth factor receptor. Neuron 1990; 4:487-492.

140. Hempstead BL, Martin-Zanca D, Kaplan DR et al. High-affinity NGF binding requires coexpression of the *trk* proto-oncogene and the low-affinity NGF receptor. Nature (Lond) 1991; 350:678-683.

141. Pioro EP, Cuello AC. Purkinje cells of adult rat cerebellum express nerve growth factor receptor immunoreactivity: Light microscopic observations. Brain Res 1988; 455:182-186.

142. Martínez-Murillo R, Caro L, Nieto-Sampedro M. Lesion-induced expression of low-affinity nerve growth factor receptor-immunoreactive protein in Purkinje cells of the adult rat. Neuroscience 1993; 52:587-593.

143. Leibrock J, Lotspeich F, Hohn A et al. Molecular cloning and expression of brain-derived neurotrophic factor. Nature (Lond) 1989; 341:149-152.

144. Martin-Zanca D, Barbacid M, Parada LF. Expression of the *trk* proto-oncogene is restricted to the sensory cranial and spinal ganglia of neural crest origin in mouse development. Genes Dev 1990; 4:683-694.

145. Squinto SP, Stitt TN, Aldrich TH et al. *Trk*B encodes a functional receptor for brain-derived neurotrophic factor and neurotrophin-3 but not nerve growth factor. Cell 1991; 65:885-893.

146. Mount HTJ, Dean DO, Alberch J et al. Glial cell line-derived neurotrophic factor promotes the survival and morphologic differentiation of Purkinje cells. Proc Natl Acad Sci USA 1995; 92:9092-9096.

CHAPTER 3

THE HUMAN
CEREBELLAR ATAXIAS

INTRODUCTION

Spinocerebellar degenerations in humans comprise a broad category of disorders affecting the cerebellum along with its afferent and efferent projections.[1-3] Many of these disorders are genetically determined. Primary foci of degeneration can be located either in the spinal cord or in the cerebellar parenchyma or both. Usually, a transsynaptic cascade of regressive events follows the primary degenerative lesion.

The term ataxia refers to a failure in muscular coordination that may be due to several related neurological disorders causing a slow, progressive deterioration of cerebellar and spinal neurons. As neurons continue to degenerate, signals to the muscles are reduced, and muscles become less and less responsive to commands from higher brain centers, thus making coordination problems more pronounced. The salient clinical features of the cerebellar ataxias are stumbling gait, hand and finger incoordination, slurred speech (dysarthria), and disturbances in eye movements.[4,5]

The age of onset, although quite variable, is usually between the second and fifth decades of life; symptoms progress, and the ability to walk may be lost within 15 years, leading to confinement to a wheelchair.[6] There is no known therapy, and ataxic disorders may be fatal, due to respiratory and cardiac complications.[4] Generally, the spinocerebellar degenerations may not affect the intellectual faculties of patients until far advanced stages.

The overall prevalence of hereditary ataxia is placed at about six cases per 100,000 population.[7,8] Higher prevalence rates have been described in isolated inbred populations in areas such as Siberia,[9] Cuba,[10,11] Cantabria, Spain[12] and Tunisia.[13] The National Ataxia Foundation estimates that about 150,000 people are affected by the hereditary ataxias and related disorders in the United States.[14] Many more, yet asymptomatic people, may be at risk. The ataxias respect no particular age, sex or race.

The classification of the various forms of the spinocerebellar ataxias is a complicated and highly debated topic. Many different

Table 3.1. Chromosomal assignment of genes causing various forms of hereditary ataxias in humans

Form of ataxia	Gene location (Hsa)
Friedreich's ataxia	9q13-q21.1
Spinocerebellar ataxia 1	6p23
Spinocerebellar atrophy II	12q24
Machado-Joseph disease/ Spinocerebellar ataxia 3	14q24.3-q31/14q24.3-qter
Spinocerebellar ataxia, type 4	16q
Spinocerebellar ataxia, type 5	11p11-q11
Infantile spinocerebellar ataxia with sensory neuropathy	10q23.3-q24.1
Cerebellar ataxia with retinal degeneration	3p21.1-p12
Ataxia-Telangiectasia	11q22.3
Ataxia with isolated vitamin E deficiency	8q13.1-q13.3
Paroxysmal acetazolamide-responsive cerebellar ataxia	19p13
Dentatorubropallidoluysian atrophy	12pter-p12

Information compiled from: Human Genetic Disorders. Index of health disorders and chromosome locations. Supplement to Journal of NIH Research, Vol 7, No 9, 1995. © 1995 The Journal of NIH Research.

schemes are used to categorize them.[8,15-20] Such schemes can be based either on the clinical features or on the pathological anatomy of the disease and take into consideration such factors as form of occurrence (genetic or sporadic), mode of inheritance (autosomal or sex-linked, recessive or dominant), molecular genetic defect, age of onset (early or late), anatomical structure predominantly affected (spinal cord or cerebellar parenchyma), cell type involved (Purkinje or granule cells), pathogenesis (primary or secondary) and involvement of extracerebellar systems (combined or multiple systems degenerations).[21-24]

Much progress has been made on the molecular genetics of the spinocerebellar ataxias and related disorders.[25] Table 3.1 gives a list of genetic diseases associated with cerebellar ataxia that have been mapped to specific chromosomes in humans. That list may be especially useful in defining mutant mouse models of human spinocerebellar disorders, considering the homology between specific human (Hsa) and mouse (Mmu) chromosomes.[26,27]

What follows is a brief description of several cerebellar degenerations and groups of neurological diseases with cerebellar involvement. Our goal here is to give an idea of the spectrum of brain disorders that affect the cerebellum and its related pathways. For a more detailed coverage of the human ataxias, the reader is referred to the extensive neurological and neuropathological literature.[1,3-6,15,17-19,28-31]

PREDOMINANTLY SPINAL DEGENERATIONS

FRIEDREICH'S ATAXIA

Friedreich's ataxia[32] is a familial form of early onset and is the most frequently encountered of all the hereditary ataxias. The disease is inherited as an autosomal recessive trait with clinical manifestation during the first two decades of life. The heterozygote frequency is 1 per 110 persons in England, while the prevalence of the clinical disease is 1-2 per 100,000 people in Europe and North America.[3,17,19]

Genetic linkage studies have mapped the genetic mutation to Hsa 9.[33] A gene, called *X25*, has been identified that encodes a

210 amino acid protein, frataxin, and the majority of Friedreich's ataxia patients studied are homozygous for an unstable GAA trinucleotide expansion repeat in the first intron of *X25*;[34] whereas normal chromosomes only have 7-22 copies of the GAA trinucleotide, affected individuals show a repetition in the order of 100-200 times.

Neuronal degeneration involves the posterior columns of the spinal cord, with the severest changes in the fasciculus gracilis, as well as Clarke's columns, the accessory nucleus cuneatus, the posterior and anterior spinocerebellar tracts, and the dorsal roots and dorsal root ganglia. In the cerebellum there is mild-to-moderate cortical degeneration and atrophy of the dentate nucleus and superior cerebellar peduncle. Outside the nervous system, one observes skeletal deformity and concentric hypertrophic cardiomyopathy.

OTHER SPINAL DEGENERATIONS

Besides Friedreich's ataxia, three additional conditions are classified among spinocerebellar ataxias with predominantly spinal degeneration.[31]

Hereditary spastic paraplegia of Strümpell[35] may be autosomal dominant, recessive or X-linked. There is axon and myelin loss in the spinocerebellar tracts and degeneration of the dorsal columns, particularly of the fasciculus gracilis. Other lesions involve cortical pyramidal neurons and the corticospinal tracts.

Hereditary posterior column ataxia of Biemond[36] is autosomal dominant. Degeneration is observed in the dorsal columns and dorsal spinal roots. The involvement of the spinocerebellar tract and of Purkinje cells is inconsistent.

Spinopontine degeneration of Boller-Segarra[37] is dominantly inherited and characterized by atrophy of Clarke's column, ventral pontine nuclei and middle cerebellar peduncle.

CEREBELLAR—BRAINSTEM DEGENERATIONS

OLIVOPONTOCEREBELLAR ATROPHY

Olivopontocerebellar atrophy is a pathological term and comprises both familial[38] and sporadic[39] cases of ataxia.[40] The inherited form is dominant. The most characteristic finding is a marked atrophy of the inferior olivary complex, the nuclei pontis and middle cerebellar peduncles. A lesser degree of atrophy is seen in the cerebellum, with Purkinje cell loss being more pronounced than granule cell loss. The posterior columns of the spinal cord and the spinocerebellar tracts are also affected, as are the corticospinal tracts and the anterior horns. In the brain, the commonest lesion is nerve cell loss and diffuse gliosis of substantia nigra and putamen. Forms of multiple systems (olivopontocerebellar and striatonigral) degeneration can be combined with autonomic nervous system involvement in what is known as Shy-Drager syndrome.[41]

DENTATORUBRAL ATROPHY

Primary atrophy of the dentate system was described by Hunt[42] in familial cases with early onset. A severe loss of neurons in the dentate nuclei results in atrophy of the superior cerebellar peduncle, through the red nucleus to the ventrolateral nucleus of thalamus.

MACHADO-JOSEPH DISEASE

An ataxic multiple systems degeneration, Machado-Joseph disease, was described in families of Azorean origin in the United States.[43] Additional cases have been reported in families from Japan, India, China, Brazil, Australia and Israel.[44] Degenerative lesions involve upper motor neurons, as well as the nigral, spinal and dentate systems. The Machado-Joseph disease/spinocerebellar ataxia 3 locus has been mapped to a region of Hsa 14 that involves an unstable trinucleotide (CAG) repeat expansion.[45] The CAG repeat length ranges from 13-41 copies on normal chromosomes and 62-80 copies on affected chromosomes;[46] the same trinucleotide repeat expansion has been found on Hsa 6 in spinocerebellar ataxia 1.

PREDOMINANTLY CEREBELLAR DEGENERATIONS

PARENCHYMAL CEREBELLAR DEGENERATION

The familial parenchymal cerebellar atrophy of Holmes[47] or cerebello-olivary degeneration is of adult onset. A similar, nonfamilial condition constitutes the late-onset cerebellar degeneration of Marie-Foix-Alajouanine.[48] Both of these conditions affect the cerebellum selectively. Histopathologically, one sees a loss of virtually all Purkinje cells, considerable loss of granule cells and a marked glial reaction in all layers of the cerebellar cortex. Such changes are accompanied by a secondary (transsynaptic) loss of inferior olivary neurons.

Both autosomal dominant and autosomal recessive forms of late-onset cerebellar parenchymal ataxias are described.[19,49] Genetic linkage studies in certain kindreds with autosomal dominant cerebellar ataxia type I point to a centromeric locus on the short arm of Hsa 6 in association with the HLA histocompatibility complex locus.[50-52]

GRANULE CELL LAYER ATROPHY

The familial cerebellar degeneration of Norman type[53] primarily affects cerebellar granule cells with a relative sparing of Purkinje cells. The atrophy of the granule cell layer has been well defined neuropathologically as a congenital disease, inherited as autosomal recessive, and characterized by ataxia, seizures, mental retardation, microcephaly and choreoathetosis.[54] Unattached spines in Purkinje dendrites, deprived of a presynaptic parallel fiber terminal element, are similar to those observed in experimental animals with granule cell degeneration induced either by X-ray irradiation[55] or neurological mutations.[56] Outside the cerebellum, loss of neurons is also seen in the nuclei of cranial nerves III and VI and in substantia nigra.[57]

ATAXIA-TELANGIECTASIA

Ataxia telangiectasia, also referred to as Louis-Bar's disease or Boder-Sedgwick syndrome, is an autosomal recessive condition with onset during the first decade of life.[31,58,59] Dysarthria and choreoathetotic movements are accompanied by the development of cutaneous telangiectasias in the conjunctivae, ear lobes and upper neck region.

The most prominent neuropathological finding in the CNS is severe loss of Purkinje cells, occasionally accompanied by posterior column degeneration, as well as neuron loss in the substantia nigra, locus coeruleus, inferior olivary complex and cranial nerve nuclei; eosinophilic inclusions are seen inside Purkinje cells and Lewy bodies inside nigral and locus coeruleus neurons.[60,61] Patients with ataxia telangiectasia have deficiencies in both humoral and cell-mediated immunity,[19] including a reduction of IgG and IgA concentrations in serum.[62]

The ataxia-telangiectasia locus has been mapped to the long arm of Hsa 11.[63] The mutated ataxia-telangiectasia gene, *ATM*, has been identified by positional cloning.[64]

Although patients homozygous for the ataxia-telangiectasia gene are numbered in the hundreds (or 1 per 40,000 births), heterozygotes are estimated to be in the 2.5 million range (i.e. 1-2% of the general population);[65] the latter incidence is particularly important in view of the fact that relatives of ataxia-telangiectasia patients have a proneness to certain forms of neoplastic disease.[66]

CONGENITAL MALFORMATIONS, METABOLIC DISORDERS AND DISEASES WITH DEFECTIVE DNA REPAIR

These conditions comprise a long list of disorders that include cerebellar involvement in their overall pathological phenotype.[3,15,17,19,67] Developmental malformations with cerebellar agenesis or dysgenesis include the Arnold-Chiari malformation, Dandy-Walker syndrome, Marinesco-Sjögren syndrome, Joubert syndrome, Gillespie syndrome and dysequilibrium syndrome. Metabolic diseases affecting the cerebellum include disorders of the urea cycle, aminoacidurias, disorders of pyruvate and lactate metabolism, storage disorders such as adrenoleukodystrophy, Niemann-Pick disease and Krabbe's disease, mitochondrial encephalomyopathies, abetalipoproteinemia, and vitamin E deficiency. Disorders associated with defective DNA repair that affect the cerebellum include xeroderma pigmentosum and Cockayne's syndrome.

PRION DISEASES

Patients with Creutzfeldt-Jakob disease[8,68,69] and Gerstmann-Sträussler-Scheinker disease[70,71] may present ataxic signs in coexistence with the progressive dementia.

NEUROTOXIC, HYPOXIC-ISCHEMIC
AND PARANEOPLASTIC ENCEPHALOPATHIES

Owing to their cellular volume size and high metabolic activity, Purkinje cells are second only to hippocampal pyramidal neurons in the hierarchy of neuronal vulnerability.[15] As a consequence, massive, diffuse involvement of Purkinje cells is seen in alcoholic cerebellar atrophy and the associated malnutrition, in phenylhydantoin intoxication and in anoxic and ischemic situations due to cardiac arrest or circulatory insufficiency.[15,72] Immunopathological paraneoplastic degenerations may affect the cerebellum and lead to Purkinje cell loss as well.[72-75]

REFERENCES

1. Hirano A, Llena JF. Degenerative diseases of the central nervous system. In: Rosenberg RN, Schochet SS Jr, eds. The Clinical Neurosciences, Vol 3: Neuropathology. New York, Edinburgh: Churchill Livingstone, 1983:285-324.

2. Barbeau A, Sadibelouiz M, Sadibelouiz A et al. A clinical classification of hereditary ataxias. Can J Neurol Sci 1984; 11:501-505.

3. Aicardi J. Heredodegenerative disorders. In: Aicardi J, ed. Diseases of the Nervous System in Childhood. London: Mac Keith Press, 1992:518-588.

4. Rosenberg RN. Hereditary ataxias. In: Rowland LP, ed. Merritt's Textbook of Neurology. Philadelphia: Lea & Febiger, 1989:627-635.

5. Plum F. Ataxia and related gait disorders. In: Wyngaarden JB, Smith LH Jr, Bennett JC, eds. Cecil Textbook of Medicine. 19th ed. Philadelphia: Saunders, 1992:2113-2115.

6. Layzer RB. Hereditary cerebellar ataxias and related disorders. In: Wyngaarden JB, Smith LH Jr, Bennett JC, eds. Cecil Textbook of Medicine. 19th ed. Philadelphia: Saunders, 1992:2138-2139.

7. Schoenberg BS. Epidemiology of the inherited ataxias. Adv Neurol 1978; 21:15-32.

8. Plaitakis A. Classification and epidemiology of cerebellar degenerations. In: Plaitakis A, ed. Cerebellar Degenerations: Clinical Neurobiology. Boston: Kluwer Academic Publishers, 1992:185-204.

9. Goldfarb LG, Chumakov MP, Petrov PA et al. Olivopontocerebellar atrophy in a large kinship in Eastern Siberia. Neurology 1989; 39:1527-1530.

10. Orosco B, Estrada R, Perry TL et al. Dominantly inherited olivopontocerebellar atrophy from eastern Cuba: Clinical, neuropathological and biochemical findings. J Neurol Sci 1989; 93:37-50.

11. Diaz GO, Fleites AN, Sagaz RC et al. Autosomal dominant cerebellar ataxia: Clinical analysis of 263 patients from a homogeneous population in Holguin, Cuba. Neurology 1990; 40:1369-1375.
12. Polo JM, Calleja J, Combarros O et al. Hereditary ataxias and paraplegias in Cantabria, Spain. An epidemiological and clinical study. Brain 1991; 114:855-866.
13. Belal S, Cancel G, Stevanin G et al. Clinical and genetic analysis of a Tunisian family with autosomal dominant cerebellar ataxia type 1 linked to the SCA2 locus. Neurology 1994; 44:1423-1426.
14. National Ataxia Foundation. Publication "Hereditary Ataxia, the Facts". Wayzata, MN, 1993.
15. Stumpf DA. Cerebellar disorders. In: Rosenberg RN, Grossman RG, eds. The Clinical Neurosciences, Vol. 2: Neurology-Neurosurgery. New York, Edinburgh: Churchill Livingstone, 1983:975-987.
16. De Negri M, Rolando S. Child ataxias: A developmental perspective. Brain Dev 1990; 195-201.
17. Harding AE. Hereditary ataxias and related disorders. In: Asbury AK, McKhann GM, McDonald WI, eds. Diseases of the Nervous System: Clinical Neurobiology. 2nd ed. Philadelphia: W B Saunders, 1992:1169-1178.
18. Harding AE. Clinical features and classification of inherited ataxias. In: Harding AE, Deufel T, eds. Inherited Ataxias. New York: Raven Press, 1993:1-14.
19. Harding AE. Neurocutaneous disorders and degenerative diseases of the spinal cord and cerebellum. In: Walton J, ed. Brain's Diseases of the Nervous System. 10th ed. Oxford: Oxford University Press, 1993:426-452.
20. Banfi S, Zoghbi HY. Molecular genetics of hereditary ataxias. Baillière's Clin Neurol 1994; 3:281-295.
21. Benomar A, Le Guern E, Durr A et al. Autosomal-dominant cerebellar ataxia with retinal degeneration (ADCA type II) is genetically different from ADCA type I. Ann Neurol 1994; 35:439-444.
22. Gouw LG, Digre KB, Harris CP et al. Autosomal dominant cerebellar ataxia with retinal degeneration: Clinical, neuropathologic, and genetic analysis of a large kindred. Neurology 1994; 44:1441-1447.
23. Enevoldson TP, Sanders MD, Harding AE. Autosomal dominant cerebellar ataxia with pigmentary macular dystrophy: A clinical and genetic study of eight families. Brain 1994; 117:445-460.
24. Iwabuchi K, Nakazawa Y, Akai J et al. Autosomal recessive hereditary cortical cerebellar atrophy with striatal degeneration: Two siblings showing choreoathetoid movement, ataxia, dementia, and amenorrhea. Brain Nerve 1994; 46:563-571.
25. Human Genetic Disorders. Index of health disorders and chromo-

some locations. Supplement to Journal of NIH Research, Vol 7, No 9, 1995.

26. Searle AG, Peters J, Lyon MF et al. Chromosome maps of man and mouse, III. Genomics 1987; 1:3-18.

27. Hillyard AL, Doolittle DP, Davisson MT et al. Locus map of mouse with comparative map points of human on mouse. Bar Harbor, ME: Jackson Laboratory, 1993.

28. Vinken PJ, Bruyn GW, De Jong JMBV, eds. Handbook of Clinical Neurology, Vol 21: System Disorders and Atrophies, Part I, chapters 14-32. Amsterdam: North-Holland Publishing Co., 1975:319-585.

29. Vinken PJ, Bruyn GW, Klawans HL et al., eds. Handbook of Clinical Neurology, Vol 60: Hereditary Neuropathies and Spinocerebellar Atrophies, chapters 21-51. Amsterdam: Elsevier Science Publishers, 1991:271-779.

30. Gilman S. Cerebellum and motor dysfunction. In: Asbury AK, McKhann GM, McDonald WI, eds. Diseases of the Nervous System: Clinical Neurobiology. 2nd ed. Philadelphia: WB Saunders, 1992:319-341.

31. Rewcastle NB. Degenerative diseases of the central nervous system. In: Davis RL, Robertson DM, eds. Textbook of Neuropathology. 2nd ed. Baltimore: Williams & Wilkins, 1991:904-961.

32. Friedreich N. Über degenerative Atrophie der spinalen Hinterstränge. Virchows Arch Pathol Anat Physiol Klin Med 1863; 26:391-419, 433-459, 27:1-26.

33. Chamberlain S, Shaw J, Rowland A et al. Mapping of mutation causing Friedreich's ataxia to human chromosome 9. Nature (Lond) 1988; 334:248-250.

34. Campuzano V, Montermini L, Moltò MD et al. Friedreich's ataxia: Autosomal recessive disease caused by an intronic GAA triplet repeat expansion. Science 1996; 271:1423-1427.

35. Strümpell A. Über eine bestimmte Form der primären kombinierten Systemerkrankung des Rückenmarks. Arch Psychiat Nervenkrank 1886; 17:217-238.

36. Biemond A. Clinisch-anatomische demonstratie over een bijzondere vorm van hereditaire ataxie. Ned Tijdschr Geneeskd 1946; 90:1014-1015.

37. Boller F, Segarra JM. Spino-pontine degeneration. Eur Neurol 1969; 2:356-373.

38. Menzel P. Beitrag zur Kenntnis der hereditären Ataxie und Kleinhirnatrophie. Arch Psychiat Nervenkrank 1891; 22:160-190.

39. Dejerine J, André-Thomas A. L' atrophie olivo-ponto-cérébelleuse. Nouv Iconogr Salpêtrière 1900; 13:330-370.

40. Berciano J. Olivopontocerebellar atrophy: A review of 117 cases. J Neurol Sci 1982; 53:253-272.
41. Shy GM, Drager GA. A neurological syndrome associated with orthostatic hypotension: A clinical-pathologic study. Arch Neurol 1960; 2:511-527.
42. Hunt JR. Dyssynergia cerebellaris myoclonica—primary atrophy of the dentate system: A contribution to the pathology and symptomatology of the cerebellum. Brain 1921; 44:490-538.
43. Nakano KK, Dawson DM, Spence A. Machado disease: A hereditary ataxia in Portuguese emigrants to Massachusetts. Neurology 1972; 22:49-55.
44. Goldberg-Stern H, D'Jaldetti R, Melamed E et al. Machado-Joseph (Azorean) disease in a Yemenite Jewish family in Israel. Neurology 1994; 44:1298-1301.
45. Takiyama Y, Oyanagi S, Kawashima S et al. A clinical and pathologic study of a large Japanese family with Machado-Joseph disease tightly linked to the DNA markers on chromosome 14q. Neurology 1994; 44:1302-1308.
46. Giunti P, Sweeney MG, Harding AE. Detection of the Machado-Joseph disease/spinocerebellar ataxia three trinucleotide repeat expansion in families with autosomal dominant motor disorders, including the Drew family of Walworth. Brain 1995; 118:1077-1085.
47. Holmes G. A form of familial degeneration of the cerebellum. Brain 1907; 30:466-489.
48. Marie P, Foix C, Alajouanine T. De l' atrophie cérébelleuse tardive à prédominance corticale. Rev Neurol 1922; 38:849-885, 1082-1111.
49. Ohta S, Mizutani Y, Anno M. An autopsy case of hereditary cerebellar atrophy (Holmes-type) with mental symptoms and rhythmic skeletal myoclonus. Brain Nerve 1994; 46:663-670.
50. Haines JL, Schut LJ, Weitkamp LR et al. Spinocerebellar ataxia in a large kindred: age at onset, reproduction, and genetic linkage studies. Neurology 1984; 34:1542-1548.
51. Kumar D, Blank CE, Gelsthorpe K. Hereditary cerebellar ataxia and genetic linkage with HLA. Hum Genet 1986; 72:327-332.
52. Zoghbi HY, Sandkuyl LA, Ott J et al. Assignment of autosomal dominant spinocerebellar ataxia (SCA1) centromeric to the HLA region on the short arm of chromosome 6, using multilocus linkage analysis. Am J Hum Genet 1989; 44:255-263.
53. Norman RM. Primary degeneration of the granular layer of the cerebellum: An unusual form of familial cerebellar atrophy occurring in early life. Brain 1940; 63:365-379.
54. Chou SM, Mizuno Y, Rothner AD. Congenital granuloprival hypoplasia of cerebellar and hippocampal cortex. J Child Neurol 1987; 2:279-286.

55. Altman J. Morphological development of the rat cerebellum and some of its mechanisms. Exp Brain Res [Suppl] 1982; 6:8-49.

56. Sotelo C. Anatomical, physiological and biochemical studies of the cerebellum from mutant mice. II. Morphological study of cerebellar cortical neurons and circuits in the weaver mouse. Brain Res 1975; 94:19-44.

57. Ferrer I, Sirvent J, Manresa JM et al. Primary degeneration of the granular layer of the cerebellum (Norman type): A Golgi study. Acta Neuropathol (Berl) 1987; 75:203-208.

58. Louis-Bar Mme. Sur un syndrome progressif comprenant des télangiectasies capillaires cutanées et conjonctivales symétriques, à disposition nævoïde et des troubles cérébelleux. Confin Neurol (Basel) 1941; 4:32-42.

59. Boder E, Sedgwick RP. Ataxia-telangiectasia: A familial syndrome of progressive cerebellar ataxia, oculocutaneous telangiectasia and frequent pulmonary infection. Pediatrics 1958; 21:526-554.

60. Strich SJ. Pathological findings in three cases of ataxia-telangiectasia. J Neurol Neurosurg Psychiat 1966; 29:489-499.

61. Monaco S, Nardelli E, Moretto G et al. Cytoskeletal pathology in ataxia-telangiectasia. Clin Neuropathol 1988; 7:44-46.

62. McFarlin DE, Strober W, Waldmann TA. Ataxia-telangiectasia. Medicine 1972; 51:281-314.

63. Gatti RA, Berkel I, Boder E et al. Localization of an ataxia-telangiectasia gene to chromosome 11q22-23. Nature (Lond) 1988; 336:577-580.

64. Savitsky K, Bar-Shira A, Gilad S et al. A single ataxia telangiectasia gene with a product similar to PI-3 kinase. Science 1995; 268:1749-1753.

65. A-T Children's Project. Publication "Accelerating Research Toward a Cure for Ataxia Telangiectasia". Boca Raton, FL, 1996.

66. Swift M, Reitnauer PJ, Morrell D et al. Breast and other cancers in families with ataxia-telangiectasia. N Engl J Med 1987; 316:1289-1294.

67. Tomiwa K, Baraitser M, Wilson J. Dominantly inherited congenital cerebellar ataxia with atrophy of the vermis. Pediatr Neurol 1987; 3:360-362.

68. Creutzfeldt HG. Über eine eigenartige herdförmige Erkrankung des Zentralnervensystems. Zbl Ges Neurol Psychiatr 1920; 57:1-18.

69. Jakob A. Über eigenartige Erkrankungen des Zentralnervensystems mit bemerkenswerten anatomischen Befunden: Spastische Pseudo-sklerose-Encephalomyelopathie mit disseminierten Degenerationsherden. Zbl Ges Neurol Psychiatr 1921; 64:147-228.

70. Gerstmann J, Sträussler E, Scheinker I. Über eine eigenartige hereditär-familiäre Erkrankung des Zentralnervensystems: Zugleich

ein Beitrag zur Frage des vorzeitigen lokalen Alterns. Zbl Ges Neurol Psychiatr 1936; 154:736-762.

71. Azzarelli B, Muller J, Ghetti B et al. Cerebellar plaques in familial Alzheimer's disease (Gerstmann-Sträussler-Scheinker variant?) Acta Neuropathol (Berl) 1985; 65:235-246.

72. Poirier J, Gray F, Escourolle R. Manual of Basic Neuropathology. Philadelphia: Saunders, 1990:151-157.

73. Greenlee JE, Brashear HR. Antibodies to cerebellar Purkinje cells in patients with paraneoplastic cerebellar degeneration and ovarian carcinoma. Ann Neurol 1983; 14:609-613.

74. Greenlee JE, Lipton HL. Anticerebellar antibodies in serum and cerebrospinal fluid of a patient with oat cell carcinoma of the lung and paraneoplastic cerebellar degeneration. Ann Neurol 1986; 19:82-85.

75. Hammack JE, Posner JB. Paraneoplastic cerebellar degeneration. In: Plaitakis A, ed. Cerebellar Degenerations: Clinical Neurobiology. Boston: Kluwer Academic Publishers, 1992:475-497.

CEREBELLAR MUTANTS
IN THE
LABORATORY MOUSE

INTRODUCTION

Mutant mice with discrete cerebellar lesions[1-3] provide invaluable experimental models of hereditary cerebellar disorders by helping us better understand developmental mechanisms of cortical histogenesis, the formation of synaptic connections and neurodegenerative processes.[4-9]

One investigative tool particularly useful in determining how the action of mutant genes brings about its effects on the affected cellular lineage is the production of chimeric mosaic aggregate organisms through the fusion of homozygous or heterozygous and wild-type embryos generally at the eight-cell stage.[10-12]

What follows is a description of the characteristic phenotypic deficits in six of the most "popular" mutants, i.e., those most often appearing in published studies in the literature. The order of description is alphabetical.

LURCHER (Lc)

The Lurcher mutation is dominant and located on mouse chromosome (Mmu) 6.[2,13-15] During the first week of life the cerebellum contains a normal number of Purkinje cells.[16] Cell loss becomes apparent at 4-6 days of age and continues during adulthood.[17] At 26 days Lurcher mutants have 90% fewer Purkinje cells than controls; that percentage becomes 97% at 63 days and 99.8% at 91 days.[18] Before their degeneration, Purkinje cells usually have more than one primary dendrite, sometimes up to five, and somatic spines persist beyond the first postnatal week.[16] Other Purkinje cell histological changes include swellings in the axons, perinuclear chromatin clumps and incomplete development of the basal polysomal mass in the soma, and hyperspinous dendrites with delayed formation of the proximal and distal compartments.[19] Moreover, Purkinje cells contain rounded mitochondria,[16] similar to those found in nervous mutant mice.[20]

Studies with explants prepared from cerebellar cortex of 2-day-old normal and Lurcher mice have revealed a significant decrease in the total area and dendritic lengths of Lurcher Purkinje cells after 15 and 20 days in vitro.[21]

Adult Lurcher mutants also show loss of granule cells.[14] Their number is reduced by 25% at four days and reaches 10% of normal by 100 days of age.[16] The molecular layer shows a substantial loss of parallel fibers after the second month of age;[22] at that stage, one also observes branching of basket cell axons in areas that had been previously occupied by Purkinje cells. In development, basket cell axons fail to develop "pinceau" formations.[19]

The inferior olivary complex is normal at eight days of age.[16] At later ages, there is a gradual cellular loss that reaches 40% at 15 days, 50% at 26 days and 75% at four months of age.[16,18] Very few climbing fibers translocate from perisomatic to peridendritic loci on the Purkinje cells.[19] Electrophysiological studies with recordings of climbing fiber responses have indicated that most Purkinje cells in developing Lurcher mutants (P14-P20) remain polyinnervated, as opposed to wild-type animals, whereby adulthood Purkinje cells are innervated by a single climbing fiber; those findings suggest that Lurcher Purkinje cells may be incompetent to respond to the granule cell input, which is thought necessary in achieving the mature monoinnervation status.[23]

To determine the primary target of the action of the *Lc* gene, investigators have used wild-type ↔ mutant

, by fusing normal and mutant embryos at the eight-cell stage; the chimeric organism contains a mosaic of genotypically normal and mutant cells.[10-12] If a mutant gene acts intrinsically within a particular cell type, then only genetically mutant cells will display the pathological phenotype in the mosaic organism.

In the case of the Lurcher mutant, studies with wild-type ↔ mutant chimeras have shown that all of the surviving Purkinje cells belong histochemically to the wild-type genotype.[24,25] On the contrary, in the chimeric mice a higher number of genetically mutant granule cells survives compared to regular (i.e. nonmosaic) Lurcher mutants, indicating that the genetically normal Purkinje cells of the mosaic organism are capable of preventing the cellular death of some genetically mutant granule cells, thus supporting the idea that the degeneration of granule cells is indirect or secondary to the action of the *Lc* gene on Purkinje cells.[26,27]

The remaining Purkinje cells of wild-type ↔ Lurcher chimeras have small somata, reduced dendritic arbors and multiple dendritic processes, as evidenced in Golgi-impregnated preparations, making them indistinguishable from Purkinje cells in Lurcher mutants.[28] Other manifestations of dendritic atrophy include a failure of the distal segment to reach the pial surface, the loss of isoplanarity and the frequent appearance of large-caliber branches terminating abruptly in "stub ends".[29] The somata of Purkinje cells in the chimeras display an increased number of lysosomes and improper configuration of the rough endoplasmic reticulum.[30]

The inferior olivary complex of the chimeras contains fewer neurons than normal mice, but more neurons than Lurcher mutant mice; histochemically one observes neurons of both genotypes.[24] Again, genetically normal Purkinje cells in the chimeras apparently prevent the degeneration of genetically mutant olivary neurons by supplying a synaptic target; thus, the cellular death of neurons in the inferior olivary complex is most likely secondary to the primary loss of Purkinje cells.[24]

The subcellular mechanisms of pathology in the Lurcher mutant are incompletely understood. Clues are provided by certain lines of research. One study, using western blotting and immunofluorescence with an antibody against a large fragment of dystrophin, a molecule normally localized along punctate foci in the plasma membrane of neuronal perikarya and dendrites but not axons, has shown absence of dystrophin expression in Lurcher cerebellum, suggesting a possible defect in plasma membrane organization, particularly as an anchor of the postsynaptic apparatus.[31] Another study examined insulin-like growth factor I (IGF-I)-dependent phosphorylation and protein tyrosine kinase activity in cerebellar cortex and found a decrease in [^{125}I]IGF-I binding in the Lurcher molecular layer, alongside a reduction in IGF-I receptor autophosphorylation, suggesting that the process of cerebellar degeneration may be associated with altered functions of the IGF-I system.[32] Lastly, light and electron microscopic studies of the pattern of Purkinje cell death in the Lurcher mutant in conjunction with in situ end-labeling and in situ hybridization methods point to morphological and histochemical features characteristic of apoptosis, including nuclear condensation, axon beading, membrane blebbing, and abnormal presence of nicked nuclear DNA and the sulfated glycoprotein 2 in Purkinje cells prior to their death.[33]

As of the writing of this chapter, the Lurcher gene has not been reported cloned. Nonetheless, intersubspecific phenotypic backcrosses have been affected and the progeny typed at seven loci in order to develop a genetic linkage map spanning about 35 cM around the *Lc* locus in Mmu 6.[15] Such a molecular linkage map represents progress toward isolating a clone of the *Lc* gene.

NERVOUS (NR)

The nervous mutation is autosomal recessive and targets Purkinje cells in the cerebellum,[34] cartwheel neurons of the dorsal cochlear nucleus[35] and retinal photoreceptors.[36] The *nr* locus maps between *D8Mit155* and *D8Mit158*, a 5.6 cM interval of Mmu 8.[37]

The majority of Purkinje cells degenerate between three and seven weeks; by two months of age, 90% of Purkinje cells in the hemispheres and 50% in the vermis have died off.[34] In the retina, 25-30% of the photoreceptors have already degenerated by three weeks; at 11 months the outer nuclear layer of the retina and the rod outer segments have disappeared.[36]

The cellular pathology of the mutation consists in a rounding of the mitochondria of Purkinje cells, initially observed at nine days of age; by 15 days all of the Purkinje cells contain rounded mitochondria.[20] The space of the Purkinje cell cytoplasm that is occupied by mitochondria is increased, and the rounded mitochondria do not migrate to the dendritic spindle.[6] Despite their unusual shape, the mitochondria retain normal histochemical activity of specific oxidative mitochondrial enzymes, such as succinic, lactic and malic dehydrogenase, the complex of NADH-Q reductase and the dehydrogenase of succinic semialdehyde.[38] The mitochondria of the cells that survive reassume normal morphology.[20]

The number of Purkinje cells is not modified significantly beyond the third month of life. Surviving Purkinje cells are arranged in wide parasagittal bands.[39] In coronal sections, Purkinje cells are disposed symmetrically to the midline along the entire rostrocaudal extent of the cerebellum. In the anterior lobe the mutation selectively affects cells that show immunopositivity with mabQ113, a monoclonal antibody that in the cerebellar cortex of rodents reveals an arrangement of Purkinje cells in parasagittal bands.

An analysis of the protein constitution of subcellular fractions of nervous cerebellum has shown that following the loss of Purkinje cells, a protein of the cytoplasmic membrane of molecular mass of about 400,000 Da, called P_{400}, is reduced substantially.[40]

The degeneration of Purkinje cells deprives parallel fibers of their postsynaptic elements shortly after normal synaptogenesis. Six to fourteen months after the degeneration of Purkinje cells, the molecular layer contains bundles of parallel fibers with numerous presynaptic boutons without a postsynaptic target. A certain proportion of such terminal boutons maintains the presynaptic membrane specialization. Quantitative analyses of the parallel fibers show that 50% of their nerve terminals are lost, whereas the remaining 50% become stabilized.[7] The persisting terminals either contact interneurons of the molecular layer or become arranged in clusters of 2-10 boutons that are surrounded by astroglial processes.[41]

The number of basket axon terminals is normal, but more than 90% of them are without a postsynaptic Purkinje cell.[41] Their ultrastructural features are in general normal; only a small proportion of basket cell axons forms heterologous synapses with dendrites of granule cells or develop electrotonic synapses with each other.[41] Immunocytochemical studies with an anti-neurofilament antibody which specifically labels basket cell axons have revealed an altered morphology of the basket cell nest in the regions devoid of Purkinje cells.[42]

The molecular layer of nervous mutant mice shows a 50% reduction in the maximum number (B_{max}) of L-[^{3}H]glutamate binding sites, whereas in the granule cell layer the B_{max} is almost unchanged; the dissociation constant (K_d) is unchanged in both layers.[43]

In the retina, the earliest cytopathological change consists in enlargement and rounding of some mitochondria in rod inner segments, with a subsequent reduction in the volume and integrity of rod outer segment membranes, an increase in the number of pyknotic photoreceptor nuclei in the outer nuclear layer and a rapid loss of photoreceptors.[44,45]

PURKINJE CELL DEGENERATION *(PCD)*

In *pcd* mutant mice there is a rapid degeneration of Purkinje cells beginning at 17 days, i.e. after the maturation of the cerebellum.[46,47] Virtually all Purkinje cells die off between the third and sixth weeks of life (Figs. 4.1 and 4.2). Purkinje cell degeneration is preceded by an abnormal retention of basal perikaryonal polyribosomes.

The *pcd* allele is recessive and has been localized to Mmu 13.[2] By screening simple sequence length polymorphisms between original and backcross strains, the *pcd* locus maps to the 3.8 cM region of Mmu 13, between *D13Mit140* and *D13Mit67*.[37] The *pcd* mutant has been used to isolate cDNA clones corresponding to mRNAs uniquely expressed by Purkinje cells.[48] Such cDNA clones have been selected from a library of normal cerebellar cDNA by virtue of their failure to hybridize to mRNA sequences from *pcd* cerebellum.

From studies employing mosaic chimeras, it appears that the site of action of the *pcd* mutant gene is intrinsic to Purkinje cells.[49] Behaviorally, homozygous *pcd* mice manifest an ataxia beginning at 3-4 weeks of age,[46] severe deficits in distal-cue spatial navigation[50] and impaired classical eyeblink conditioning.[51] Extracerebellar components of the *pcd* phenotype include degeneration of retinal photoreceptors,[36,52-54] olfactory mitral cells[55] and ventromedial geniculate thalamic neurons.[56,57]

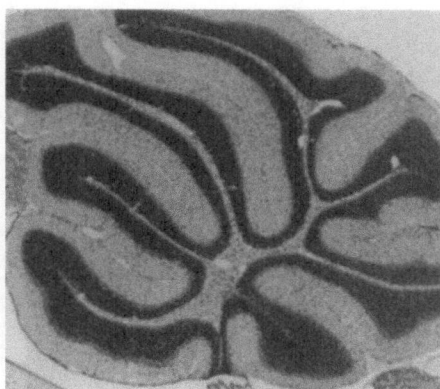
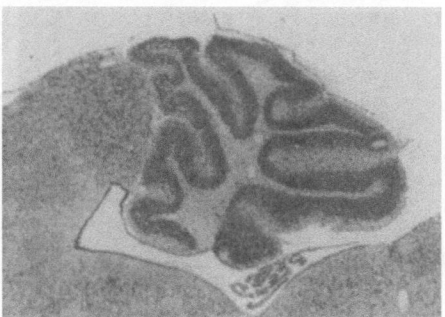
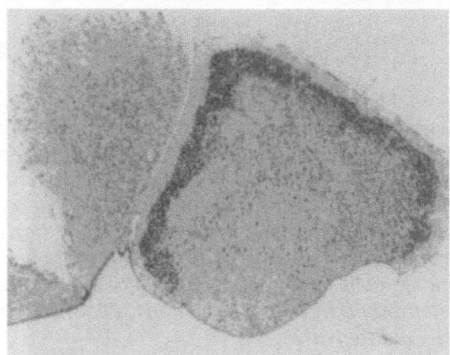
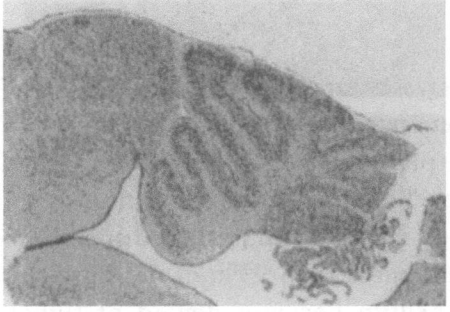

Fig. 4.1. Sagittal sections of cerebellum in wild-type (upper), Purkinje cell degeneration or pcd (second from top), reeler (third from top) and weaver mutant mice (lower). Magnification ×19. Reprinted with permission from: Kambouris M, Sangameswaran L, Dlouhy SR et al. Mol Brain Res 1993; 18:321-328. © 1993 Elsevier Science Publishers B.V.

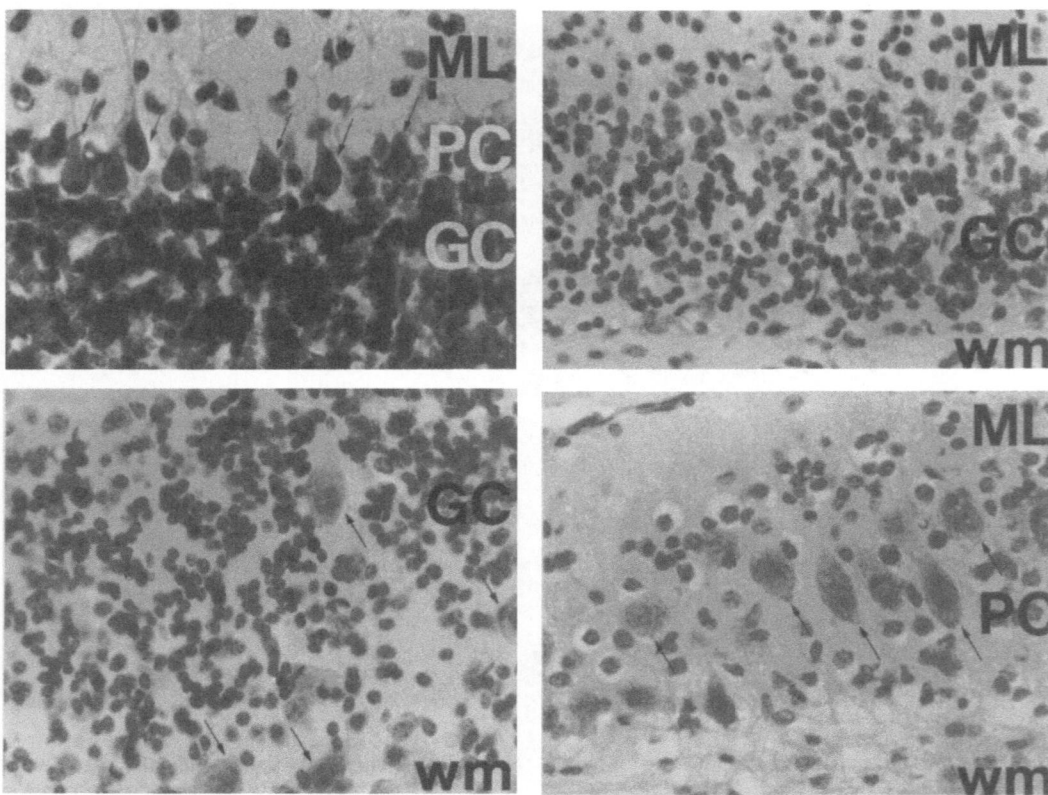

Fig. 4.2. Microscopic view of the cerebellar cortex in wild-type (upper left), Purkinje cell degeneration or pcd (upper right), reeler (lower left) and weaver (lower right) mutant mice. Histological Nissl stain. In the wild-type (upper left), the molecular layer (ML) and the layers of Purkinje cells (PC) and granule cells (GC) are clearly discernible. Arrows point at Purkinje cell somata. The dendrites of Purkinje cells can be seen in the molecular layer. In pcd cerebellum (upper right), Purkinje cells are missing, but there is a clear separation into molecular and granule cell layers. In the reeler mutant (lower left), Purkinje cells (arrows) are interspersed in the granule cell layer and in the subcortical white matter (wm). In weaver cerebellum (lower right), most of the granule cells are missing and Purkinje cells (arrows) are not aligned in a monolayer. Magnification x320. Reprinted with permission from: Kambouris M, Sangameswaran L, Dlouhy SR et al. Mol Brain Res 1993; 18:321-328. © 1993 Elsevier Science Publishers B.V.

The loss of Purkinje cells progresses rapidly, such that ~25-50% of Purkinje cells have already degenerated by postnatal days 22-24.[46] The loss of Purkinje cells takes place in clusters, in such a way that, in coronal sections of cerebellum, remaining Purkinje cells are organized in parasagittal bands with a symmetrical disposition relative to the midline.[39] Eventually, more than 99% of the Purkinje cells disappear, and the *pcd* cerebellum becomes devoid of signal for specific Purkinje cell markers, such as 28 kDa Ca^{2+}-binding protein, polypeptide PEP-19, cyclic GMP-dependent protein kinase, ganglioside G_{D1a} and IGF-I.[39,58-62] In histological preparations, one can see cell debris throughout the molecular, Purkinje and granule cell layers, as well as in the subcortical white matter and the deep cerebellar nuclei.

The degeneration of Purkinje cells in the mutants leads to a loss of presynaptic afferents to the deep cerebellar nuclei and a resulting atrophy in all subdivisions, i.e., the nucleus lateralis, nucleus interpositus and nucleus medialis, which is more evident in the dorsoventral plane.[63] In the rostrocaudal plane, not much change is seen relative to the control structures. In mutants, the total volume of the deep nuclei is reduced by 22% between 23 days and 10 months of age. There is a 21% cell loss in 300-day-old mutants with respect to 23-day-old mutants. Nonetheless, the majority of neurons in the deep cerebellar nuclei (>70%) survive, even in the deafferented state.

Purkinje axon terminals are lost from the dorsal part of the lateral vestibular nucleus (Deiters' nucleus), to which Purkinje cells normally supply an axonal innervation in addition to the deep cerebellar nuclei.[64,65]

The loss of Purkinje cells also deprives inferior olivary neurons of their major postsynaptic target. At 17 days of age, the inferior olivary complex of *pcd* mutant mice does not differ in cell number from control mice. However, inferior olivary neurons in 23-day-old mutants are 23% fewer than in age-matched controls, and in 300-day-old mutants they are 48% fewer than in controls.[66] Interestingly, cell number in the inferior olive appears to become stabilized at that point, such that the amount of cell loss at 450 days of age is also ~50%.[67] The medial accessory olive appears to be less affected than the principal and dorsal accessory olives. It appears, therefore, that in the mature olivocerebellar system, the wellness of inferior olivary neurons depends on the state of their postsynaptic Purkinje cells. On the other hand, once a critical neuronal mass has degenerated in the inferior olive, the remaining neurons become stabilized, and no further loss is observed even at a very advanced age. It is conceivable that surviving neurons of the inferior olivary complex are sustained through synaptic connections of climbing fiber collaterals with cerebellar cortical interneurons or with neurons of the deep cerebellar nuclei or both.

Modifications in GABA receptors in *pcd* mutants have been studied via [^{3}H]muscimol and [^{3}H]flunitrazepam binding.[68-72] The GABA agonist muscimol binds to high-affinity GABA binding sites located on the β-subunit of the GABA$_A$ receptor complex. Central benzodiazepine binding sites are located on the α-subunit of the GABA$_A$ receptor-Cl$^-$ channel macromolecular complex. In normal mouse cerebellum, the highest concentration of [^{3}H]muscimol binding sites is in the granule cell layer; a much lower grain density is present over the Purkinje and molecular layers, and negligible numbers are over the deep nuclei and white matter. In 2-month-old *pcd* mutants, a 29% decrease in grain density is observed over the granule cell layer, while labeling is still present in the molecular layer. Neurochemical studies reveal a 50% decrease in benzodiazepine receptors in 45-day-old *pcd* mutants, i.e., after the degeneration of Purkinje cells. By 10 months of age, the decrease in benzodiazepine receptors reaches 80%.

To determine the histological localization of these receptor changes, in vitro autoradiographic techniques have been used to explore [^{3}H]flunitrazepam binding.[68,69,71] The α$_1$-subunit of the GABA$_A$/benzodiazepine receptor complex accounts for >80% of cerebellar [^{3}H]flunitrazepam binding sites, with expression in the deep cerebellar nuclei, which receive a GABAergic inhibitory innervation from Purkinje cells. The highest concentration of [^{3}H]flunitrazepam binding sites is over the molecular layer; intermediate grain density is seen over the Purkinje cell layer and intermediate-to-high density over the deep nuclei; labeling over the granule cell layer is low. In *pcd* mutants, decreases in [^{3}H]flunitrazepam were observed in the cerebellar cortex in association with the loss of Purkinje cells, and an increase in the density of benzodiazepine binding sites was in the deep cerebellar nuclei.[71] When a correction factor was introduced for layer atrophy,[73] there was a 60% decrease in benzodiazepine binding at P45 in the *pcd* molecular layer and a 84% decrease at P300. The initial decrease in benzodiazepine receptors at P45 is associated with the selective loss of Purkinje cells. The amount of receptor binding that persists at P300 reflects binding in parallel fibers of the surviving granule cells. The increase observed over the *pcd* nuclear complex probably reflects a denervation-induced supersensitivity in the deep cerebellar nuclei subsequent to the loss of innervation by Purkinje axon terminals.

Using in situ hybridization histochemistry, it was found that despite the loss of Purkinje cells, virtually all large neurons of the *pcd* deep cerebellar nuclei expressed the α_1, β_2 and γ_2 GABA$_A$ receptor subunit mRNAs; moreover, while the levels of β_2 and γ_2 subunit mRNAs were comparable between mutants and controls, the α_1 subunit mRNA levels were 40% lower than the control at P24, i.e., prior to the completion of Purkinje cell degeneration, and subsequently increased between P60-P90, possibly owing to the deafferentation, thus indicating a selective modulation of the α_1 subunit gene expression by the Purkinje cell input.[74]

Quantitative autoradiography of [^{3}H]CNQX (6-cyano-7-nitro-quinoxaline-2,3-dione) binding has been used to analyze the distribution of non-*N*-methyl-D-aspartate (non-NMDA) receptors in the cerebellum of wild-type mice and *pcd* mutants.[75] The K_d did not differ between the two conditions, but the B_{max} was reduced by 50% in the mutants. By incubating tissue sections with 200 nM [^{3}H]CNQX in vitro, a 35% decrease was observed in non-NMDA receptors in the *pcd* molecular layer, a 36% decrease in the granule cell layer, and a 33% decrease in deep cerebellar nuclei. The reduction of non-NMDA receptors in the *pcd* molecular layer supports their localization on Purkinje dendrites.

REELER *(RL)*

Reeler mutant mice (gene symbol *rl*, located on Mmu 5) have a systematic malposition of neuron classes in the cerebellum and forebrain.[2,5,76-79] Abnormalities have been described in the architectonics and development of the facial nerve nucleus as well.[80] Reeler mutant mice manifest alterations in cellular immunity, involving defective functions of T lymphocytes and macrophages.[81]

In the cerebellum (Figs. 4.1 and 4.2), the typical organization and lamination of the cortex is altered.[4,77,79] Cerebellar abnormalities are evident before birth, when Purkinje cells fail to become condensed in a clearly defined plate, and the cerebellar foliation pattern becomes increasingly deficient.[82] The salient architectonic perturbation of the postnatal reeler cerebellum consists of a malposition of Purkinje cells, most of which lie heterotopically within or below the granule cell layer.[83] The reeler cerebellum contains about 60% fewer Purkinje cells than normal, based on counts of cGMP-dependent protein kinase immunoreactive neurons.[84] Despite their ectopic location in large subcortical masses, reeler Purkinje cells show a tendency to distribute into alternating zones that become apparent after immunolabeling with either Zebrin II or P-path monoclonal antibodies; however, in reeler mutants there are an estimated 7-9 such zones as opposed to 14 major divisions in the wild-type cerebellum.[85]

The reeler cerebellum is small in size and contains a reduced complement of granule cells.[77,79] In some cases, granule cells have no T-bifurcation of their axon and assume a bipolar form.[86] In vitro studies with microexplant cultures from early postnatal cerebellum indicate that the behavior of granule cells and their precursors of reeler origin are qualitatively similar to that of normal counterparts, but the number of granule cell precursor cells in the external germinal layer may be reduced in the reeler cerebellum.[87]

Changes in the pattern of expression of specific molecular markers in the reeler cerebellum, including polypeptide PEP-19, protein L7 and clone GCAP-8 mRNA, reflect the anatomical characteristics of the mutant phenotype.[88-90]

Alterations in [^{3}H]GABA binding include a decreased affinity of the Na^+-independent high-affinity GABA-binding component of synaptosomal membranes and an increased affinity of the Na^+-dependent, high-affinity binding component in reeler cerebellar homogenates and synaptic membranes, whereas the number of either the Na^+-independent or the Na^+-dependent binding sites is not altered.[91] In situ hybridization studies with [^{35}S]cRNA probes indicate that the expression by Purkinje cells of RNA message for the α_1 and γ_2 subunits of the $GABA_A$/benzodiazepine receptor is not impaired despite their malpositioning.[72,92] A biochemical dissociation has been described concerning the expression of *N*-methyl-D-aspartate receptor channel mRNAs, as all reeler Purkinje cells express the ζ_1 subunit, only a subset expresses the σ_1 subunit, and none of them express the σ_2, σ_3 or σ_4 subunits.[93]

The morphogenetic development of the inferior olivary complex proceeds abnormally from E15 onward, leading to a cytoarchitectonically dysplastic olive.[94] It has been proposed that such an abnormal development of the reeler inferior olivary complex may depend on a local, intrinsic action of the *rl* gene on the positioning of neurons in the olivary primordium.[94]

Studies with reeler ↔ wild-type chimeras have found genetically normal Purkinje cells to be aberrantly positioned, and conversely, some reeler Purkinje cells to be located in the normal Purkinje cell layer; these data were taken to suggest that Purkinje cells in the reeler mutant are not positioned according to their own genetic information, but rather by factors extrinsic to Purkinje cells[10] and to a disturbance of neuronal migration attributable to an abnormal cell-to-cell interaction between young migrating neurons and radial glial cells.[95]

In cerebral cortex, the laminar segregation of postmigratory cells proceeds abnormally during development, and in the adult the polymorphous cell population is superficial rather than deep to the pyramidal cell classes.[5] The plexiform layer is absent, and large pyramidal neurons are located superficially, whereas medium and small-sized pyramidal cells are concentrated in the depth of the cortex, and small stellate cells are intercalated between large and medium-sized pyramidal cells. There is a general delay in the appearance of immunoreactivities for cell adhesion molecules L1, J1, N-CAM and their shared carbohydrate L2 in the embryonic cortex of reeler mutants.[96] In the visual cortex of the occipital lobe, acetylcholinesterase histochemistry reveals a distinct mosaic after the second postnatal week, perhaps in association with eye opening, as opposed to the normal pattern within bands or laminae.[97]

In Ammon's horn, the pyramidal cells are radially dispersed rather than forming a compact layer, and in the fascia dentata, granule cells are mixed with neurons of the hilus.[78] Electrophysiological studies indicate an impairment of long-term potentiation in the superficial layer of CA1 pyramidal neurons, which appears to be due mostly to strong inhibitory inputs to this malpositioned population of neurons.[98]

In molecular biological studies, one allele (*rl* [rg]), which had been isolated and characterized previously through insertional mutagenesis, was used to clone the *reelin* gene that has been deleted from two reeler alleles.[99,100] Normal, but not mutant, mice express the *reelin* gene in embryonic and postnatal neurons during the stage of neuronal migration. The encoded protein resembles proteins of the extracellular matrix that participate in processes of cellular adhesion. Thus, it appears that the reeler phenotype reflects a failure of primordial events related to layer formation in the brain that the reelin protein normally regulates.

Comparative studies reveal some similarities in early cortical organization between reeler and reptilian, particularly chelonian, embryos, most notably the presence of an inverted gradient of cortical histogenesis; in that context, the characterization and cloning of the *rl* gene may prove crucial in understanding the progressive acquisition of cortical architectonics during brain evolution and the importance of cell-cell interactions.[101]

STAGGERER (SG)

The staggerer mutation (gene symbol *sg*) has been traditionally considered as autosomal recessive, although certain abnormalities are observed in heterozygosity as well.[102,103] The *sg* locus is on Mmu 9.[3] Although a synteny exists in the region close to the centromere of human chromosome (Hsa) 11, where the genes for human spinocerebellar ataxia type 5 and ataxia-telangiectasia have been mapped, and Mmu 9, where the staggerer mutation is located,[104] it is now clear that these are different genes.

In staggerer homozygotes the external germinal layer, although atrophic, produces postmitotic neurons that migrate and differentiate normally; however, mature granule cells progressively degenerate mostly during the third and fourth weeks of life.[105-109] Purkinje cells are reduced in number by 80-90% and have abnormal dendritic branches from which the peripheral elements, i.e., the spiny branchlets, are missing.[110-114] That leads to an inability of synapse formation between the terminal boutons of the parallel fibers and the Purkinje dendritic spines, owing exactly to the absence of a postsynaptic element. Thus, despite the migration of postmitotic granule cells and the outgrowth of axons that become differentiated into parallel fibers with typical varicosities and synaptic vesicles, the process of synaptogenesis fails because of the absence of postsynaptic structures. Some parallel fiber nerve terminals form "pseudocontacts" with astrocytic membranes, suggesting that they maintain the ability for autonomous development of the presynaptic boutons. However, during the second month of age all of these presynaptic endings undergo degeneration leading to a retrograde reaction of the perikaryon and extensive necrosis of granule cells.

In association with the loss of Purkinje cells in staggerer cerebellum, there is a marked reduction in the intrinsic expression levels of the metabotropic glutamate receptor type 1α (mGluR1α), which is normally expressed in adult Purkinje cells.[115] On the other hand, concomitantly with the degeneration of granule cells, there is a progressive decline of the neuronal form of the c-*src* proto-oncogene-encoded protein-tyrosine kinase pp60$^{c\text{-}src\,(+)}$.[116] Both the molecular and granule cell layers of staggerer mutant mice show a 60% reduction in the B_{max} of L-[^{3}H]glutamate binding sites, but no significant changes of the K_d in either layer.[117]

The developmental pattern of polyinnervation of Purkinje cells by climbing fibers persists in adult staggerer mice, as parallel fibers, which are believed to provide the signal for transition to a monoinnervation stage, are missing.[118,119] A loss of neurons in the inferior olivary complex takes place in staggerer mice after birth, which is believed to be secondary (transsynaptic) to the degeneration of Purkinje cells.[120-122]

Biochemical studies have shown that in staggerer mutant mice there is a disturbance in the transition of the embryonic form of the neural cell adhesion molecule (N-CAM) to the adult form.[123] The *Ncam* gene maps to Mmu 9, but at some genetic distance from the *sg* locus.[124] On the other hand, granule cell precursors in the external germinal layer express TAG-1, an axonal glycoprotein involved in adhesion and neurite outgrowth, without any apparent impairment, suggesting that the transient expression of TAG-1 may be maintained as long as granule cell precursors are in a postmitotic, premigratory state, even if such a state is prolonged in the case of staggerer mutants.[125]

A 47 kDa polypeptide, termed SP47, appears in the mature staggerer cerebellum, but not in the wild-type.[126] Other biochemical changes in the staggerer cerebellum include elevations of the basal mRNA levels of the inflammatory cytokines interleukin 1β and interleukin 6, along with an increase in the ratio of the Kunitz protease inhibitor-bearing isoforms of amyloid β-precursor protein (βAPP) over isoform $βAPP_{695}$ compared to normal.[127]

Pathological expressions outside the cerebellum involve a delay in the development of the thymus, a general enlargement of lymph nodes and undersized spleen.[128] Furthermore, a prolonged expression of embryonic cell surface phenotypes is observed on the surface of the cerebellum, thymus and, to some extent, spleen, but not in cells of other organs.[128]

Golgi studies with staggerer ↔ wild-type chimeras have disclosed that the dendritic abnormalities induced in Purkinje cells by the *sg* mutation and consisting of rudimentary, unbranched dendrites that lack tertiary branchlet spines, reflect cell autonomous, developmental genetic blocks in the cytological maturation of cerebellar Purkinje cells.[28]

The molecular basis of the staggerer effects on cerebellar development has been unraveled.[129] The *sg* was genetically mapped to a 160 kb interval on Mmu 9 which also contained the gene encoding for RORα, a member of the nuclear hormone-receptor superfamily. Staggerer mice carry a deletion within the *RORα* gene, which prevents translation of the ligand-binding homology domain. Based on these findings, a model has been proposed whereby RORα interacts with the thyroid hormone signaling pathway to induce Purkinje cell maturation.[129]

WEAVER *(wv)*

The weaver *(wv)* mutation occurred spontaneously in a C57BL/6J mouse genetic stock in 1961[130] and has been shown to interfere with neuronal survival in the cerebellar cortex and in the mesotelencephalic dopamine (DA) projection system.[1,131,132] Traditionally, the weaver mutation has been considered autosomal recessive, based on its effects on locomotor behavior.[1] However, the partial expression of cerebellar deficits in heterozygosity points rather to an incomplete dominant trait.[7,133,134]

The cerebellum of homozygous weaver mutants is grossly atrophic.[4,135] Despite normal mitotic activity,[133] the majority of postmitotic granule cell precursors in the external germinal layer (EGL) do not emit axons, fail to migrate inward to the internal granule cell layer and die massively at the interface of the EGL and the molecular layer during the first two weeks of postnatal life.[135] The residual EGL persists in weaver mutants for about one week longer than in control mice.[1] The adult cerebellum of weaver mutants appears agranular (Figs. 4.1 and 4.2) with the exception of the most lateral portions of the hemispheres and the paraflocculus, where granule cell loss is less pronounced.[1]

Weaver heterozygotes (*wv*/+) show a reduced rate of granule cell migration and an intermediate degree of granule cell death;[133,135] arrested granule cells are seen at the interface of the molecular layer and Purkinje cell layer of adult animals.

Evidence of cell death in the postmitotic zone of the EGL is already apparent at birth, indicating that an earlier event in granule cell development, such as the exit of neuroblasts from the cell cycle or axonogenesis, is affected by the *wv* gene.[136] Delaying EGL cells from exiting the cell cycle by hormonal manipulation reduces granule cell death in both *wv*/*wv* and *wv*/+, thus supporting the notion that granule cell degeneration occurs after cell division has been completed.[137]

In parallel with granule cell loss, the levels of several molecules associated with granule cells in cerebellum decline postnatally. Such molecules include sulfonylurea receptors associated with ATP-regulated K⁺ channels,[138] the ATP-dependent glutamate uptake system in synaptic vesicles,[139] the neuronal form of the *c-src* protooncogene-encoded protein-tyrosine kinase pp60$^{c-src(+)}$,[116,140] saxitoxin-sensitive Na⁺ channels[141] and receptors for ω-conotoxin GVIA, an irreversible blocker of N-type Ca^{2+} channel,[142] which are all thought to be positioned presynaptically on parallel fibers. Decreases in the granule cell-specific protein β2 chimerin and in calcicludine high-affinity binding sites, as well as in the Purkinje cell-specific protein Wnt-3, have been reported in the weaver cerebellum.[143-145] The cellular localization of growth-associated phosphoprotein GAP-43, microtubule-associated protein MAP2 and βAPP mRNAs correlates with the corresponding anatomical deficits.[146,147]

Purkinje cell counts in midsagittal sections of cerebellum have disclosed a reduction of Purkinje cell number, evident as early as postnatal day 5, both in heterozygosity and homozygosity, with little change occurring thereafter.[133] The average reduction in vermal

Purkinje cell number is 48% in homozygous weaver mutants and 20% in weaver heterozygotes.[133] These counts have been confirmed independently in three- to nine-months-old mice, in which the vermal Purkinje cell deficit was estimated to be 50% and 21% in the two genotypes, respectively.[148] The overall reduction in Purkinje cell number is 28% in *wv/wv* and 14% in *wv/+*,[148] indicating that Purkinje cell loss is severer in vermis than in the hemispheres.

The remaining Purkinje cells of the *wv/wv* cerebellar cortex do not form a monolayer and are found in ectopic positions.[135,149] The apical dendrites of Purkinje cells are oriented randomly, are often inversed, and occasionally display a "weeping willow" shape of arborization.[110,149,150] Further, Purkinje cell dendrites do not develop spiny branchlets, while their primary and secondary branches have an irregular rough surface owing to the presence of numerous spines.[151] Purkinje cell arrangement in more than one row and aberrant dendritic trees of lesser severity are seen in the heterozygote cerebellum.

At the ultrastructural level, one finds unattached Purkinje cell dendritic spines, often complete with postsynaptic specialization densities, and usually devoid of a presynaptic input from parallel fibers; structurally, dendritic spines themselves are indistinguishable from those in normal cerebellar cortex with a presynaptic element.[135,151-153] The synaptic remodeling of neuronal circuitry in the weaver cerebellum involves the formation of heterologous synapses on free spines by axon terminals of mossy fiber rosettes and by climbing fiber varicosities, which never reach such targets under normal circumstances, and the formation of attachment plate-like junctions by free postsynaptic sites of apposing Purkinje dendritic spines.[151]

A major afferent input to Purkinje cells comes from olivo-cerebellar climbing fibers. In normal development, there is a transient multiple innervation of Purkinje cells by climbing fibers, which later regresses, reaching monoinnervation in the adult; parallel fiber input to Purkinje cells is thought to be involved in the regression of supernumerary climbing fiber collaterals.[154] Since granule cells degenerate in the weaver cerebellum, the multiple innervation of Purkinje cells by climbing fibers persists in adult mutants at a mean ratio of 3.5 between the two neuronal elements.[119,154,155]

Midbrain DA neurons constitute an additional cellular target of the *wv* mutation. DA cell loss is found in substantia nigra (area A9), ventral tegmental area (area A10) and retrorubral nucleus (area A8).[156] Studies combining [³H]thymidine dating and TyrOHase immunocytochemistry indicate that the mutation preferentially kills DA neurons that are generated later in embryonic life.[157]

Neuron losses in the mesencephalic DA cell groups lead to DA deficiency in several telencephalic areas to which A8, A9 and A10 normally supply DA innervation, including caudate-putamen complex, olfactory tubercle, frontal cortex and lateral septal nucleus.[131,158,159] Quantitative autoradiographic studies of [³H]DA uptake by striatal slices have disclosed severer deficits in DA uptake and storage compared to DA levels, even in areas like the nucleus accumbens, where DA content is normal.[160] Adding a further degree of complexity to the functional deficit, remaining DA axon terminals in the weaver neostriatum establish an inadequate synaptic connectivity with resident striatal neurons.[161]

Heterozygous weaver mice have comparable to normal nigrostriatal DA cell somata and axons, but they display a severe defect in the dendritic DA projection that extends from the pars compacta to the pars reticulata of the substantia nigra.[132,162,163] Moreover, the remaining dendrites have thinner-than-normal varicosities and receive a decreased percentage of synaptic input from striatonigral axon terminals.[164]

Behaviorally, *wv/wv* mice manifest instability of gait, poor limb coordination and resting and intention tremors,[1] in all likelihood underlain by the cerebellar and nigrostriatal pathologies.[165] On the other hand, *wv/+* animals do not present with either locomotor abnormalities or tremor, but intermittently manifest generalized tonic/clonic convulsions, which are usually lethal.[166,167] One logical explanation for such an activity could pertain to the DAergic dendrite deficit, as the substantia nigra has been implicated in the pathophysiology of experimental seizures.[168,169]

Studies employing +/+ ↔ *wv*/+ chimeras have indicated that the *wv* gene intrinsically affects granule cells in causing them to be ectopic.[170] Additional studies with intraspecific and interspecies +/+ ↔ *wv*/*wv* chimeras led to the conclusion that granule cell death is most likely due to an intrinsic action of the *wv* gene on granule cells.[171] Moreover, the decrease in Purkinje cell number seems to be a direct effect of the mutation.[134]

The *wv* allele has been mapped to the distal end of Mmu 16,[172,173] within a phylogenetically conserved region highly homologous to telomeric Hsa 21.[174] The *wv* mutation has been identified as a missense mutation with a G to A substitution in nucleotide 953 of the inward rectifier K[+] channel gene *Girk2* and an ensuing Gly to Ser replacement at residue 156 of the GIRK2 protein.[175] There is a human equivalent ATP-sensitive K[+] channel gene that maps to Hsa 21q22.1.[176] Initial electrophysiological experiments to assess the functional implications of the Gly to Ser substitution have not demonstrated inward rectifying K[+] currents in cultured granule cells from either wild-type or weaver mice at postnatal day eight.[177] Subsequent analyses of GIRK2 weaver and GIRK1 channels in Xenopus oocytes found that GIRK2 weaver homomultimeric channels lose their selectivity for K[+] ions, giving rise to inappropriate receptor-activated and basally active Na[+] currents, while heteromultimers of GIRK2 weaver and GIRK1 appeared to have reduced current.[178]

Cultured weaver granule cells are proteolytically overactive and secrete excessive amounts of tissue plasminogen activator, which is likely to interfere with neurite outgrowth potential on a laminin substratum.[179] Electrophysiology shows that weaver granule cells have poor resting membrane potentials (-38 mV); aprotinin, a protease inhibitor, can restore the resting membrane potential to near normal (-59 mV), rescue weaver granule cells from death on the laminin substratum and promote neurite outgrowth.[179]

OTHER MUTATIONS

Several other mutations lead to different types of cerebellar malformation or degeneration in the laboratory mouse, thus providing additional ataxic models.[3] Among these other cerebellar mutants, one may list the dreher mutant mouse (*dr/dr*, Mmu 1),[180-182] the dystonia musculorum *(dt/dt,* Mmu 1),[183-187] the hot-foot mutant *(ho/ho,* Mmu 6),[188] the hyperspiny Purkinje cell *(hpc/hpc),*[189-191] the meander tail *(mea/mea,* Mmu 4),[145,192-196] the quivering *(qv/qv,* Mmu 7),[197-200] the rostral cerebellar malformation *(rcm/rcm,* Mmu 3),[201] the stumbler *(stu/stu),*[202-204] the swaying *(sw/sw,* Mmu 15)[205] and the tambaleante mutant mouse *(tbl/tbl).*[39,206]

There is a good chance that transgenic[207,208] or knockout[209] cerebellar mouse models may find use in neural transplantation studies. In contrast to the "randomness" of the spontaneous mutations, these other models could in theory be directed to bear more resemblance to specific human diseases.

REFERENCES

1. Sidman RL, Green MC, Appel SH. Catalog of the Neurological Mutants of the Mouse. Cambridge, MA: Harvard University Press, 1965.
2. Lyon MF, Searle AG, eds. Genetic Variants and Strains of the Laboratory Mouse. 2nd ed. Oxford, Stuttgart: Oxford University Press, Gustav Fischer Verlag, 1989.
3. Catalog of Neurological Mouse Models Available from the Jackson Laboratory. Bar Harbor, ME, November 1995.
4. Sidman RL. Development of interneuronal connections in brains of mutant mice. In: Carlson FD, ed. Physiological and Biochemical Aspects of Nervous Integration. Englewood Cliffs, NJ: Prentice Hall, 1968:163-193.
5. Caviness VS, Rakic P. Mechanisms of cortical development: A view from mutations in mice. Ann Rev Neurosci 1978; 1:297-326.
6. Landis DMD, Landis SC. Several mutations in mice that affect the cerebellum. Adv Neurol 1978; 21:85-105.
7. Sotelo C. Mutant mice and the formation of cerebellar circuitry. Trends Neurosci 1980; 3:33-36.
8. Sidman RL. Mutations affecting the central nervous system in the mouse. In: Schmitt FO, Bird SJ, Bloom FE, eds. Molecular Genetic Neuroscience. New York: Raven Press, 1982:389-400.
9. Goldowitz D, Eisenman LM. Genetic mutations affecting murine cerebellar structure and function. In: Driscoll P, ed. Genetically Defined Animal Models of Neurobehavioral Dysfunctions. Boston-Basel-Berlin: Birkhäuser, 1992:66-88.
10. Mullen RJ. Genetic dissection of the CNS with mutant-normal mouse and rat chimeras. In Cowan WM, Ferrendelli JA, eds. Society for Neuroscience Symposia, Vol II: Approaches to the Cell Biology of Neurons. Bethesda: Society for Neuroscience, 1977:47-65.

11. Mullen RJ. Mosaicism in the central nervous system of mouse chimeras. In: Subtelny S, Sussex IM, eds. The Clonal Basis of Development. New York: Academic Press, 1978:83-101.

12. Mullen RJ, Herrup K. Chimeric analysis of mouse cerebellar mutants. In: Breakfield XO, ed. Neurogenetics: Genetic Approaches to the Nervous System. Amsterdam: Elsevier/North-Holland, 1979:173-196.

13. Phillips RJS. "Lurcher", a new gene in linkage group XI of the house mouse. J Genet 1960; 57:35-42.

14. Caddy KWT, Biscoe TJ. Preliminary observations on the cerebellum in the mutant mouse Lurcher. Brain Res 1975; 91:276-280.

15. Norman DJ, Fletcher C, Heintz N. Genetic mapping of the Lurcher locus on mouse chromosome 6 using an intersubspecific backcross. Genomics 1991; 9:147-153.

16. Caddy KWT, Biscoe TJ. Structural and quantitative studies on the normal C3H and Lurcher mutant mouse. Phil Trans Roy Soc Lond (Biol) 1979; 287:167-201.

17. Swisher DA, Wilson DB. Cerebellar histogenesis in the Lurcher *(Lc)* mutant mouse. J Comp Neurol 1977; 173:205-218.

18. Caddy KWT, Biscoe TJ. The number of Purkinje cells and olive neurones in the normal and Lurcher mutant mouse. Brain Res 1976; 111: 396-398.

19. Dumesnil-Bousez N, Sotelo C. Early development of the Lurcher cerebellum: Purkinje cell alterations and impairment of synaptogenesis. J Neurocytol 1992; 21:506-529.

20. Landis SC. Ultrastructural changes in the mitochondria of cerebellar Purkinje cells of nervous mutant mice. J Cell Biol 1973; 57:782-797.

21. Doughty ML, Patterson L, Caddy KWT. Cerebellar Purkinje cells from the Lurcher mutant and wild-type mouse grown in vitro: A light and electron microscope study. J Comp Neurol 1995; 357:161-179.

22. Wilson DB. Histological defects in the cerebellum of adult Lurcher *(Lc)* mice. J Neuropathol Exp Neurol 1976; 35:40-45.

23. Rabacchi SA, Bailly Y, Delhaye-Bouchaud N et al. Role of the target in synapse elimination: Studies in cerebellum of developing Lurcher mutants and adult chimeric mice. J Neurosci 1992; 12:4712-4720.

24. Wetts R, Herrup K. Interaction of granule, Purkinje and inferior olivary neurons in Lurcher chimeric mice. I. Qualitative studies. J Embryol Exp Morphol 1982; 68:87-98.

25. Wetts R, Herrup K. Cerebellar Purkinje cells are descended from a small number of progenitors committed during early development: Quantitative analysis of Lurcher chimeric mice. J Neurosci 1982; 2:1494-1498.

26. Wetts R, Herrup K. Interaction of granule, Purkinje and inferior olivary neurons in Lurcher chimeric mice. II. Granule cell death. Brain Res 1982; 250:358-362.

27. Wetts R, Herrup K. Direct correlation between Purkinje and granule cell number in the cerebella of Lurcher chimeras and wild-type mice. Dev Brain Res 1983; 10:41-47.

28. Soha JM, Herrup K. Stunted morphologies of cerebellar Purkinje cells in Lurcher and staggerer mice are cell-intrinsic effects of the mutant genes. J Comp Neurol 1995; 357:65-75.

29. Caddy KWT, Herrup K. Studies of the dendritic tree of wild-type cerebellar Purkinje cells in Lurcher chimeric mice. J Comp Neurol 1990; 297:121-131.

30. Caddy KWT, Herrup K. The fine structure of the Purkinje cell and its afferents in Lurcher chimeric mice. J Comp Neurol 1991; 305:421-434.

31. Lidov HGW, Byers TJ, Kunkel LM. The distribution of dystrophin in the murine central nervous system: An immunocytochemical study. Neuroscience 1993; 54:167-187.

32. Vig PJS, Desaiah D, Joshi P et al. Decreased insulin-like growth factor I-mediated protein tyrosine phosphorylation in human olivopontocerebellar atrophy and Lurcher mutant mouse. J Neurol Sci 1994; 124:38-44.

33. Norman DJ, Feng L, Cheng SS et al. The Lurcher gene induces apoptotic death in cerebellar Purkinje cells. Development 1995; 121:1183-1193.

34. Sidman RL, Green MC. Nervous, a new mutant mouse with cerebellar disease. In: Sabourdy M, ed. Les mutants pathologiques chez l' animal. Paris: Éditions du Centre National de la Recherche Scientifique, 1970:69-79.

35. Berrebi AS, Mugnaini E. Effects of the murine mutation 'nervous' on neurons in cerebellum and dorsal cochlear nucleus. J Neurocytol 1988; 17:465-484.

36. Mullen RJ, LaVail MM. Two new types of retinal degeneration in cerebellar mutant mice. Nature (Lond) 1975; 258:528-530.

37. Campbell DB, Hess EJ. Chromosome localization of the neurological mouse mutations tottering *(tg)*, Purkinje cell degeneration *(pcd)*, and nervous *(nr)*. Soc Neurosci Abstr 1995; 21:2110.

38. Landis SC. Histochemical demonstration of mitochondrial dehydrogenases in developing normal and nervous mutant mouse Purkinje cells. J Histochem Cytochem 1975; 23:136-143.

39. Wassef M, Sotelo C, Cholley B, Brehier A, Thomasset M. Cerebellar mutations affecting the postnatal survival of Purkinje cells in the mouse disclose a longitudinal pattern of differentially sensitive cells. Dev Biol 1987; 124:379-389.

40. Mallet J, Huchet M, Pougeois R et al. Anatomical, physiological and biochemical studies on the cerebellum from mutant mice. III. Protein differences associated with the weaver, staggerer and nervous mutations. Brain Res 1976; 103:291-312.

41. Sotelo C, Triller A. Fate of presynaptic afferents to Purkinje cells in the adult nervous mutant mouse: A model to study presynaptic stabilization. Brain Res 1979; 175:11-36.

42. Brion JP, Guilleminot J, Nunez J. Dendritic and axonal distribution of the microtubule-associated proteins MAP2 and tau in the cerebellum of the nervous mutant mouse. Dev Brain Res 1988; 44:221-232.

43. Angelatou F, Mitsacos A, Gouras V et al. L-aspartate and L-glutamate binding sites in developing normal and 'nervous' mutant mouse cerebellum. Int J Dev Neurosci 1987; 5:373-381.

44. LaVail MM, White MP, Gorrin GM et al. Retinal degeneration in the nervous mutant mouse. I. Light microscopic cytopathology and changes in the interphotoreceptor matrix. J Comp Neurol 1993; 333:168-181.

45. White MP, Gorrin GM, Mullen RJ et al. Retinal degeneration in the nervous mutant mouse. II. Electron microscopic analysis. J Comp Neurol 1993; 333:182-198.

46. Mullen RJ, Eicher EM, Sidman RL. Purkinje cell degeneration, a new neurological mutation in the mouse. Proc Natl Acad Sci USA 1976;73:208-212.

47. Landis SC, Mullen RJ. The development and degeneration of Purkinje cells in *pcd* mutant mice. J Comp Neurol 1978; 177:125-144.

48. Nordquist DT, Kozak CA, Orr HT. cDNA cloning and characterization of three genes uniquely expressed in cerebellum by Purkinje neurons. J Neurosci 1988; 8:4780-4789.

49. Mullen RJ. Site of *pcd* gene action and Purkinje cell mosaicism in the cerebella of chimeric mice. Nature (Lond) 1977; 270:245-247.

50. Goodlett CR, Hamre KM, West JR. Dissociation of spatial navigation and visual guidance performance in Purkinje cell degeneration *(pcd)* mutant mice. Behav Brain Res 1992; 47:129-141.

51. Chen L, Bao S, Kim JJ et al. Impaired classical eyeblink conditioning in Purkinje cell degeneration *(pcd)* mutant mice. Soc Neurosci Abstr 1995; 21:1221.

52. LaVail MM, Blanks JC, Mullen RJ. Retinal degeneration in the *pcd* cerebellar mutant mouse. I. Light microscopic and autoradiographic analysis. J Comp Neurol 1982; 212:217-230.

53. Blanks JC, Mullen RJ, LaVail MM. Retinal degeneration in the *pcd* cerebellar mutant mouse. II. Electron microscopic analysis. J Comp Neurol 1982; 212:231-246.

54. Blanks JC, Spee C. Retinal degeneration in the *pcd/pcd* mutant mouse: Accumulation of spherules in the interphotoreceptor space. Exp Eye Res 1992; 54:637-644.

55. Greer CA, Shepherd GM. Mitral cell degeneration and sensory function in the neurological mutant mouse Purkinje cell degeneration *(pcd)*. Brain Res 1982; 235:156-161.

56. O'Gorman S, Sidman RL. Degeneration of thalamic neurons in 'Purkinje cell degeneration' mutant mice. I. Distribution of neuron loss. J Comp Neurol 1985; 234:277-297.

57. O'Gorman S. Degeneration of thalamic neurons in 'Purkinje cell degeneration' mutant mice. II. Cytology of neuron loss. J Comp Neurol 1985; 234:298-316.

58. Sotelo C, Alvarado-Mallart RM. Cerebellar transplantations in adult mice with heredo-degenerative ataxia. Ann NY Acad Sci 1987; 495:242-267.

59. Chang AC, Triarhou LC, Alyea CJ et al. Developmental expression of polypeptide PEP-19 in cerebellar suspensions transplanted into the cerebellum of *pcd* mutant mice. Exp Brain Res 1989; 76:639-645.

60. Ghetti B, Triarhou LC. The Purkinje cell degeneration mutant: A model to study the consequences of neuronal degeneration. In: Plaitakis A, ed. Cerebellar Degenerations: Clinical Neurobiology. Boston: Kluwer Academic; 1992:159-181.

61. Furuya S, Irie F, Hashikawa T et al. Ganglioside $G_{D1\alpha}$ in cerebellar Purkinje cells: Its specific absence in mouse mutants with Purkinje cell abnormality and altered immunoreactivity in response to conjunctive stimuli causing long term desensitization. J Biol Chem 1994; 269:32418-32425.

62. Zhang W, Lee W-H, Triarhou LC. Grafted cerebellar cells in a mouse model of hereditary ataxia express IGF-I system genes and partially restore behavioral function. Nature Med 1996; 2:65-71.

63. Triarhou LC, Norton J, Ghetti B. Anterograde transsynaptic degeneration in the deep cerebellar nuclei of Purkinje cell degeneration *(pcd)* mutant mice. Exp Brain Res 1987; 66:577-588.

64. Bäurle J, Grover BG, Grüsser-Cornehls U. Plasticity of GABAergic terminals in Deiters' nucleus of weaver mutant and normal mice: A quantitative light microscopic study. Brain Res 1992; 591: 305-318.

65. Bäurle J, Grüsser-Cornehls U. Calbindin D-28k in the lateral vestibular nucleus of mutant mice as a tool to reveal Purkinje cell plasticity. Neurosci Lett 1994; 167:85-88.

66. Triarhou LC, Norton J, Ghetti B. Morphometric analysis of the inferior olivary complex in *pcd* mutant mice. Neurosci Lett 1986; [Suppl] 26:111.

67. Triarhou LC, Ghetti B. Stabilisation of neurone number in the inferior olivary complex of aged 'Purkinje cell degeneration' mutant mice. Acta Neuropathol (Berl) 1991; 81:597-602.

68. Rotter A, Frostholm A. Cerebellar benzodiazepine receptor distribution: An autoradiographic study of the normal C57BL/6J and Purkinje cell degeneration mutant mouse. Neurosci Lett 1986; 71:66-71.

69. Rotter A, Frostholm A. Cerebellar benzodiazepine receptors: Cellular localization and consequences of neurological mutations in mice. Brain Res 1988; 444:133-146.

70. Rotter A, Gorenstein C, Frostholm A. The localization of GABA$_A$ receptors in mice with mutations affecting the structure and connectivity of the cerebellum. Brain Res 1988; 439:236-248.

71. Kahle G, Kaulen P, Bruning G et al. Autoradiographic analysis of benzodiazepine receptors in mutant mice with cerebellar defects. J Chem Neuroanat 1990; 3:261-270.

72. Luntz-Leybman V, Frostholm A, Fernando L et al. GABA$_A$/benzodiazepine receptor γ_2 subunit gene expression in developing normal and mutant mouse cerebellum. Mol Brain Res 1993; 19:9-21.

73. Vaccarino FM, Ghetti B, Nurnberger JI. Residual benzodiazepine binding in the cortex of *pcd* mutant cerebella and qualitative binding in the deep cerebellar nuclei of control and mutant mice: An autoradiographic study. Brain Res 1985; 343:70-78.

74. Gambarana C, Loria CJ, Siegel RE. GABA$_A$ receptor messenger RNA expression in the deep cerebellar nuclei of Purkinje cell degeneration mutants is maintained following the loss of innervating Purkinje neurons. Neuroscience 1993; 52:63-71.

75. Stasi K, Mitsacos A, Triarhou LC et al. Functional integration of transplanted Purkinje cells into the atrophic cerebellum: I. Excitatory amino acid receptors and afferent innervation. Abstr Am Soc Neural Transpl 1996; 3:50.

76. Falconer DS. Two new mutants, "trembler" and "reeler", with neurological actions in the house mouse. J Genet 1951; 50:192-201.

77. Hamburgh M. Observations on the neuropathology of 'reeler', a neurological mutation in mice. Experientia (Basel) 1960; 16: 460-461.

78. Caviness VS, Sidman RL. Retrohippocampal, hippocampal, and related structures of the forebrain in the reeler mutant mouse. J Comp Neurol 1973; 147:235-254.

79. Mariani J, Crepel F, Mikoshiba K et al. Anatomical, physiological and biochemical studies of the cerebellum from reeler mutant mouse. Phil Trans Roy Soc Lond (Biol) 1977; 281:1-28.

80. Goffinet AM. Abnormal development of the facial nerve nucleus in reeler mutant mice. J Anat 1984; 138:207-215.

81. Green-Johnson JM, Zalcman S, Vriend CY et al. Suppressed T cell and macrophage function in the 'reeler' *(rl/rl)* mutant, a murine strain with elevated cerebellar norepinephrine concentration. Brain Behav Immun 1995; 9:47-60.

82. Goffinet AM. The embryonic development of the cerebellum in normal and reeler mutant mice. Anat Embryol (Berl) 1983; 168:73-86.

83. Goffinet AM, So KF, Yamamoto M et al. Architectonic and hodological organization of the cerebellum in reeler mutant mice. Dev Brain Res 1984; 16:263-276.

84. Heckroth JA, Goldowitz D, Eisenman LM. Purkinje cell reduction in the reeler mutant mouse: A quantitative immunohistochemical study. J Comp Neurol 1989; 279:546-555.

85. Edwards MA, Leclerc N, Crandall JE et al. Purkinje cell compartments in the reeler mutant mouse as revealed by Zebrin II and 90-acetylated glycolipid antigen expression. Anat Embryol (Berl) 1994; 190:417-428.

86. Terashima T, Inoue K, Inoue Y et al. Observations on Golgi epithelial cells and granule cells in the cerebellum of the reeler mutant mouse. Dev Brain Res 1985; 18:103-112.

87. Nagata I, Terashima T. Migration behaviour of granule cells on laminin in cerebellar microexplant cultures from early postnatal reeler mutant mice. Int J Dev Neurosci 1994; 12:387-395.

88. Sangameswaran L, Hempstead J, Morgan JI. Molecular cloning of a neuron-specific transcript and its regulation during normal and aberrant cerebellar development. Proc Natl Acad Sci USA 1989; 86:5651-5655.

89. Smeyne RJ, Oberdick J, Schilling K et al. Dynamic organization of developing Purkinje cells revealed by transgene expression. Science 1991; 254:719-721.

90. Kambouris M, Sangameswaran L, Dlouhy SR et al. Cellular distribution of the RNA transcripts of a newly discovered gene in the brain of normal, weaver, Purkinje cell degeneration and reeler mutant mice as evidenced by in situ hybridization histochemistry. Mol Brain Res 1993; 18:321-328.

91. Matsokis N, Valcana T. [^{3}H]GABA binding in the cerebellum of the reeler murine mutant. Neurochem Int 1985; 7:37-44.

92. Frostholm A, Zdilar D, Chang A et al. Stability of GABA$_A$/benzodiazepine receptor α_1 subunit mRNA expression in reeler mouse cerebellar Purkinje cells during postnatal development. Dev Brain Res 1991; 64:121-128.

93. Watanabe M, Nakagawa S, Takayama C et al. Cerebellum of the adult reeler mutant mouse contains two Purkinje cell populations with respect to gene expression for the *N*-methyl-D-aspartate receptor channel. Neurosci Res 1995; 22:335-345.

94. Goffinet AM. The embryonic development of the inferior olivary complex in normal and reeler (rl^{ORL}) mutant mice. J Comp Neurol 1983; 219:10-24.

95. Terashima T, Inoue K, Inoue Y et al. Observations on the cerebellum of normal ↔ reeler mutant mouse chimera. J Comp Neurol 1986; 252:264-278.

96. Godfraind C, Schachner M, Goffinet AM. Immunohistological localization of cell adhesion molecules L1, J1, N-CAM and their common carbohydrate L2 in the embryonic cortex of normal and reeler mice. Dev Brain Res 1988; 42:99-111.

97. Steindler DA, Faissner A, Harrington KL. A unique mosaic in the visual cortex of the reeler mutant mouse. Cerebral Cortex 1994; 4:129-137.

98. Ishida A, Shimazaki K, Terashima T et al. An electrophysiological and immunohistochemical study of the hippocampus of the reeler mutant mouse. Brain Res 1994; 662:60-68.

99. Miao GG, Smeyne RJ, D'Arcangelo G et al. Isolation of an allele of reeler by insertional mutagenesis. Proc Natl Acad Sci USA 1994; 91:11050-11054.

100. D'Arcangelo G, Miao GG, Chen S-C et al. A protein related to extracellular matrix proteins deleted in the mouse mutant *reeler.* Nature (Lond) 1995; 374:719-723.

101. Goffinet AM. The reeler gene: A clue to brain development and evolution. Int J Dev Biol 1992; 36:101-107.

102. Shojaeian H, Delhaye-Bouchaud N, Mariani J. Decreased number of cells in the inferior olivary nucleus of the adult mouse (+/*sg*) heterozygous for the staggerer gene. Neuroscience 1987; 22:91-97.

103. Bakalian A, Kopmels B, Messer A et al. Peripheral macrophage abnormalities in mutant mice with spinocerebellar degeneration. Res Immunol 1992; 143:129-139.

104. Gatti RA, Berkel I, Boder E et al. Localization of an ataxia-telangiectasia gene to chromosome 11q22-23. Nature (Lond) 1988; 336:577-580.

105. Sidman RL, Lane PW, Dickie MM. Staggerer, a new mutation in the mouse affecting the cerebellum. Science 1962; 137:610-612.

106. Sax DS, Hirano A, Shofer RJ. Staggerer, a neurological murine mutant: An electron microscopic study of the cerebellar cortex in the adult. Neurology 1968; 18:1093-1100.

107. Yoon CH. Developmental mechanism for changes in cerebellum of 'staggerer' mouse, a neurological mutant of genetic origin. Neurology 1972; 22:743-754.

108. Sotelo C, Changeux J-P. Transsynaptic degeneration 'en cascade' in the cerebellar cortex of staggerer mutant mice. Brain Res 1974; 67:519-526.

109. Hirano A, Dembitzer HM. The fine structure of staggerer cerebellum. J Neuropathol Exp Neurol 1975; 34:1-11.
110. Sotelo C. Dendritic abnormalities of Purkinje cells in the cerebellum of neurologic mutant mice (weaver and staggerer). Adv Neurol 1975; 12:335-351.
111. Yoon CH. Pleiotropic effect of the staggerer gene. Brain Res 1976; 109:206-215.
112. Landis DMD, Reese TS. Structure of the Purkinje cell membrane in staggerer and weaver mutant mice. J Comp Neurol 1977; 171:247-260.
113. Landis DMD, Sidman RL. Electron microscopic analysis of postnatal histogenesis in the cerebellar cortex of staggerer mutant mice. J Comp Neurol 1978; 179:831-863.
114. Bradley P, Berry M. The Purkinje cell dendritic tree in mutant mouse cerebellum: A quantitative Golgi study of weaver and staggerer mice. Brain Res 1978; 142:135-141.
115. Ryo Y, Miyawaki A, Furuichi T et al. Expression of the metabotropic glutamate receptor mGluR1 alpha and the ionotropic glutamate receptor GluR1 in the brain during the postnatal development of normal mouse and in the cerebellum from mutant mice. J Neurosci Res 1993; 36:19-32.
116. Wiestler OD, Trenkner E, Walter G. Progressive loss of neuronal src protein in postnatal weaver and staggerer cerebellum. Exp Cell Biol 1988; 56:190-195.
117. Kouvelas ED, Mitsacos A, Angelatou F et al. Glutamate receptors in mammalian cerebellum: Alterations in human ataxic disorders and cerebellar mutant mice. In: Plaitakis A, ed. Cerebellar Degenerations: Clinical Neurobiology. Boston: Kluwer Academic Publishers, 1992:123-137.
118. Crepel F, Mariani J. Anatomical, physiological and biochemical studies of the cerebellum from mutant mice. I. Electrophysiological analysis of cerebellar cortical neurons in the staggerer mouse. Brain Res 1975; 98:135-147.
119. Mariani J. Extent of multiple innervation of Purkinje cells by climbing fibers in the olivocerebellar system of weaver, reeler, and staggerer mutant mice. J Neurobiol 1982; 13:119-126.
120. Blatt GJ, Eisenman LM. A qualitative and quantitative light microscopic study of the inferior olivary complex in the adult staggerer mutant mouse. J Neurogenet 1985; 2:51-66.
121. Shojaeian H, Delhaye-Bouchaud N, Mariani J. Decreased number of cells in the inferior olivary nucleus of the developing staggerer mouse. Dev Brain Res 1985; 21:141-146.
122. Shojaeian-Zanjani H, Herrup K, Guastavino JM et al. Developmental studies of the inferior olivary nucleus in staggerer mutant mice. Dev Brain Res 1994; 82:18-28.

123. Edelman GM, Chuong CM. Embryonic to adult conversion of neural cell adhesion molecules in normal and staggerer mice. Proc Natl Acad Sci USA 1982; 79:7036-7040.

124. D'Eustachio P, Davisson MT. Resolution of the staggerer *(sg)* mutation from the neural cell adhesion molecule locus *(Ncam)* on mouse Chromosome 9. Mamm Genome 1993; 4:278-280.

125. Karagogeos D, Kyriakopoulou K, Delhaye-Bouchaud N et al. Cerebellar granule cell differentiation in mutant and X-irradiated rodents as revealed by the neural adhesion molecule TAG-1. Soc Neurosci Abstr 1995; 21:1036.

126. Heinlein UAO, Ruppert C, Wille W. Staggerer-specific protein SP47: A unique species among age- and genotype-dependent cerebellar proteins. Neurochem Res 1987; 12:53-60.

127. Brugg B, Dubreuil YL, Huber G et al. Inflammatory processes induce β-amyloid precursor protein changes in mouse brain. Proc Natl Acad Sci USA 1995; 92:3032-3035.

128. Trenkner E, Hoffmann MK. Defective development of the thymus and immunological abnormalities in the neurological mouse mutation 'staggerer'. J Neurosci 1986; 6:1733-1737.

129. Hamilton BA, Frankel WN, Kerrebrock AW et al. Disruption of the nuclear hormone receptor RORα in staggerer mice. Nature (Lond) 1996; 379:736-739.

130. Lane PW. Mouse News Lett 1964; 30:32.

131. Lane JD, Nadi NS, McBride WJ et al. Contents of serotonin, norepinephrine and dopamine in the cerebrum of the 'staggerer', 'weaver' and 'nervous' neurologically mutant mice. J Neurochem 1977; 29:349-350.

132. Triarhou LC. Weaver gene expression in central nervous system. In: Conn PM, ed. Gene Expression in Neural Tissues. San Diego: Academic Press, 1992:209-227.

133. Rezai Z, Yoon CH. Abnormal rate of granule cell migration in the cerebellum of 'weaver' mutant mice. Dev Biol 1972; 29:17-26.

134. Smeyne RJ, Goldowitz D. Purkinje cell loss is due to a direct action of the weaver gene in Purkinje cells: Evidence from chimeric mice. Dev Brain Res 1990; 52:211-218.

135. Rakic P, Sidman RL. Sequence of developmental abnormalities leading to granule cell deficit in cerebellar cortex of weaver mutant mice. J Comp Neurol 1973; 152:103-132.

136. Smeyne RJ, Goldowitz D. Development and death of external granular layer cells in the weaver mouse cerebellum: A quantitative study. J Neurosci 1989; 9:1608-1620.

137. Smeyne RJ, Goldowitz D. Postnatal development of the wild-type and weaver cerebellum after embryonic administration of propylthiouracil (PTU). Dev Brain Res 1990; 54:282-286.

138. Mourre C, Widmann C, Lazdunski M. Sulfonylurea binding sites associated with ATP-regulated K⁺ channels in the central nervous system: Autoradiographic analysis of their distribution and ontogenesis, and of their localization in mutant mice cerebellum. Brain Res 1990; 519:29-43.

139. Fischer-Bovenkerk C, Kish PE, Ueda T. ATP-dependent glutamate uptake into synaptic vesicles from cerebellar mutant mice. J Neurochem 1988; 51:1054-1059.

140. Brugge JS, Lustig A, Messer A. Changes in the pattern of expression of pp60*c-src* in cerebellar mutants of mice. J Neurosci Res 1987; 18:532-538.

141. Mourre C, Widmann C, Lazdunski M. Saxitoxin-sensitive Na⁺ channels: Presynaptic localization in cerebellum and hippocampus of neurological mutant mice. Brain Res 1990; 533:196-202.

142. Maeda N, Wada K, Yuzaki M et al. Autoradiographic visualization of a calcium channel antagonist, [¹²⁵I]ω-conotoxin GVIA, binding site in the brains of normal and cerebellar mutant mice *(pcd* and *weaver)*. Brain Res 1989; 489:21-30.

143. Leung T, How BE, Manser E et al. Cerebellar β2-chimaerin, a GTPase-activating protein for p21 Ras-related Rac is specifically expressed in granule cells and has a unique *N*-terminal SH2 domain. J Biol Chem 1994; 269:12888-12892.

144. Schweitz H, Heurteaux C, Bois P et al. Calcicludine, a venom peptide of the Kunitz-type protease inhibitor family, is a potent blocker of high-threshold Ca²⁺ channels with a high affinity for L-type channels in cerebellar granule neurons. Proc Natl Acad Sci USA 1994; 91:878-882.

145. Salinas PC, Copeland NG, Jenkins NA et al. Maintenance of Wnt-3 expression in Purkinje cells of the mouse cerebellum depends on interactions with granule cells. Development 1994; 120:1277-1286.

146. Solà C, Mengod G, Palacios JM et al. GAP-43 and MAP2 expression in normal and weaver cerebellum: Immunohistochemical and in situ hybridization studies. Brain Pathol 1994; 4:395.

147. Solà C, Mengod G, Ghetti B et al. Regional distribution of the alternatively spliced isoforms of βAPP RNA transcript in the brain of normal, heterozygous and homozygous weaver mutant mice as revealed by in situ hybridization histochemistry. Mol Brain Res 1993; 17:340-346.

148. Blatt GJ, Eisenman LM. A qualitative and quantitative light microscopic study of the inferior olivary complex of normal, reeler, and weaver mutant mice. J Comp Neurol 1985; 232:117-128.

149. Rakic P, Sidman RL. Organization of cerebellar cortex secondary to deficit of granule cells in weaver mutant mice. J Comp Neurol 1973; 152:133-162.

150. Sotelo C. Purkinje cell ontogeny: Formation and maintenance of spines. Prog Brain Res 1978; 48:149-170.

151. Sotelo C. Anatomical, physiological and biochemical studies of the cerebellum from mutant mice. II. Morphological study of cerebellar cortical neurons and circuits in the weaver mouse. Brain Res 1975; 94:19-44.

152. Hirano A, Dembitzer HM. Cerebellar alterations in the weaver mouse. J Cell Biol 1973; 56:478-486.

153. Hanna RB, Hirano A, Pappas GD. Membrane specializations of dendritic spines and glia in the weaver mouse cerebellum: A freeze-fracture study. J Cell Biol 1976; 68:403-410.

154. Crepel F, Mariani J. Multiple innervation of Purkinje cells by climbing fibers in the cerebellum of the weaver mutant mouse. J Neurobiol 1976; 7:579-582.

155. Puro DG, Woodward DJ. The climbing fiber system in the weaver mutant. Brain Res 1977; 129:141-146.

156. Triarhou LC, Norton J, Ghetti B. Mesencephalic dopamine cell deficit involves areas A8, A9 and A10 in weaver mutant mice. Exp Brain Res 1988; 70:256-265.

157. Bayer SA, Wills KV, Triarhou LC et al. Selective vulnerability of late-generated dopaminergic neurons of the substantia nigra in weaver mutant mice. Proc Natl Acad Sci USA 1995; 92:9137-9140.

158. Schmidt MJ, Sawyer BD, Perry KW et al. Dopamine deficiency in the weaver mutant mouse. J Neurosci 1982; 2:376-380.

159. Roffler-Tarlov S, Graybiel AM. Weaver mutation has differential effects on the dopamine-containing innervation of the limbic and non-limbic striatum. Nature (Lond) 1984; 307:62-66.

160. Doucet G, Brundin P, Seth S et al. Degeneration and graft-induced restoration of dopamine innervation in the weaver mouse neostriatum: A quantitative radioautographic study of [^{3}H]dopamine uptake. Exp Brain Res 1989; 77:552-568.

161. Triarhou LC, Norton J, Ghetti B. Synaptic connectivity of tyrosine hydroxylase immunoreactive nerve terminals in the striatum of normal, heterozygous and homozygous weaver mutant mice. J Neurocytol 1988; 17:221-232.

162. Triarhou LC. Definition of the Mesostriatal Dopamine Deficit in the Weaver Mutant Mouse and Reconstruction of the Damaged Pathway by Means of Neural Transplantation. Ann Arbor, MI: University Microfilms International, 1987.

163. Triarhou LC, Ghetti B. The dendritic dopamine projection of the substantia nigra: Phenotypic denominator of weaver gene action in hetero- and homozygosity. Brain Res 1989; 501:373-381.

164. Triarhou LC, Ghetti B. Further characterization of the dopaminergic dendrite deficit in substantia nigra pars reticulata of

heterozygous and homozygous weaver mutant mice: Golgi, MAP2 and synaptic connectivity studies. Soc Neurosci Abstr 1991; 17:159.

165. Triarhou LC, Ghetti B. Neuroanatomical substrate of behavioural impairment in weaver mutant mice. Exp Brain Res 1987; 68:434-435.

166. Seyfried TN. Convulsive disorders. In: Foster HL, Small JD, Fox JG, eds. The Mouse in Biomedical Research. New York: Academic Press, 1982; 4:97:124.

167. Eisenberg B, Messer A. Tonic/clonic seizures in a mouse mutant carrying the weaver gene. Neurosci Lett 1989; 96: 168-172.

168. Lindvall O, Ingvar M, Gage FH. Short term status epilepticus in rats causes specific behavioral impairments related to substantia nigra necrosis. Exp Brain Res 1986; 64:143-148.

169. La Grutta V, Sabatino M. Substantia nigra-mediated anticonvulsant action: A possible role of a dopaminergic component. Brain Res 1990; 515:87-93.

170. Goldowitz D, Mullen RJ. Granule cell as a site of gene action in the weaver mouse cerebellum: Evidence from heterozygous mutant chimeras. J Neurosci 1982; 2:1474-1485.

171. Goldowitz D. The weaver granuloprival phenotype is due to intrinsic action of the mutant locus in granule cells: Evidence from homozygous weaver chimeras. Neuron 1989; 2:1565-1575.

172. Lane PW, Sweet HO. Mouse News Lett 1979; 60:46,50.

173. Mjaatvedt AE, Citron MP, Reeves RH. High-resolution mapping of *D16Led-1*, *Gart*, *Gas-4*, *Cbr*, *Pcp-4*, and *Erg* on distal mouse chromosome 16. Genomics 1993; 17:382-386.

174. Reeves RH, Crowley MR, Lorenzon N et al. The mouse neurological mutant *weaver* maps within the region of chromosome 16 that is homologous to human chromosome 21. Genomics 1989; 5:522-526.

175. Patil N, Cox DR, Bhat D et al. A potassium channel mutation in weaver mice implicates membrane excitability in granule cell differentiation. Nature Genet 1995; 11:126-129.

176. Tsaur M-L, Menzel S, Lai F-P et al. Isolation of a cDNA clone encoding a K_{ATP} channel-like protein expressed in insulin-secreting cells, localization of the human gene to chromosome band 21q22.1, and linkage studies with NIDDM. Diabetes 1995; 44:592-596.

177. Mjaatvedt AE, Cabin DE, Cole SE et al. Assessment of a mutation in the H5 domain of *Girk2* as a candidate for the weaver mutation. Genome Res 1995; 5:453-463.

178. Slesinger PA, Patil N, Liao J et al. Functional effects of the mouse *weaver* mutation on G protein-gated inwardly rectifying K^+ channels. Neuron 1996; 16:321-331.

179. Murtomaki S, Trenkner E, Wright JM et al. Increased proteolytic activity of the granule neurons may contribute to neuronal death in the weaver mouse cerebellum. Dev Biol 1995; 168:635-648.

180. Sekiguchi M, Shimai K, Guo H et al. Cytoarchitectonic abnormalities in hippocampal formation and cerebellum of dreher mutant mouse. Dev Brain Res 1992; 67:105-112.

181. Sekiguchi M, Nowakowski RS, Shimai K et al. Abnormal distribution of acetylcholinesterase activity in the hippocampal formation of the dreher mutant mouse. Brain Res 1993; 622:203-210.

182. Sekiguchi M, Abe H, Shimai K et al. Disruption of neuronal migration in the neocortex of the dreher mutant mouse. Dev Brain Res 1994; 77:37-43.

183. Duchen LW, Strich SJ, Falconer DS. Clinical and pathological studies of an hereditary neuropathy in mice *(dystonia musculorum)*. Brain 1964; 87:367-378.

184. Janota I. Ultrastructural studies of an hereditary sensory neuropathy in mice *(dystonia musculorum)*. Brain 1972; 95:529-536.

185. Duchen LW. Dystonia musculorum—an inherited disease of the nervous system in the mouse. Adv Neurol 1976; 14:353-365.

186. Ebendal T, Lundin L-G. Nerve growth factor in three neurologically deficient mouse mutants. Neurosci Lett 1984; 50:121-126.

187. Sotelo C, Guenet JL. Pathologic changes in the CNS of *dystonia musculorum* mutant mouse: An animal model for human spinocerebellar ataxia. Neuroscience 1988; 27:403-424.

188. Guastavino JM, Sotelo C, Damez-Kinselle I. Hot-foot murine mutation: behavioral effects and neuroanatomical alterations. Brain Res 1990; 523:199-210.

189. Guenet JL, Sotelo C, Mariani J. Hyperspiny Purkinje cell, a new neurological mutation in the mouse. J Hered 1983; 74:105-108.

190. Sotelo C. Axonal abnormalities in cerebellar Purkinje cells of the 'hyperspiny Purkinje cell' mutant mouse. J Neurocytol 1990; 19:737-755.

191. Frederic F, Hainaut F, Thomasset M et al. Cell counts of Purkinje and inferior olivary neurons in the 'hyperspiny Purkinje cell' mutant mouse. Eur J Neurosci 1992; 4:127-135.

192. Ross ME, Fletcher C, Mason CA et al. Meander tail reveals a discrete developmental unit in the mouse cerebellum. Proc Natl Acad Sci USA 1990; 87:4189-4192.

193. Fletcher C, Norman DJ, Heintz N. Genetic mapping of meander tail, a mouse mutation affecting cerebellar development. Genomics 1991; 9:647-655.

194. Eisenman LM, Arlinghaus LE. Spinocerebellar projection in the meander tail mutant mouse: Organization in the granular posterior lobe and the agranular anterior lobe. Brain Res 1991; 558:149-152.

195. Eisenman LM, Pruett JR. Expression of the Purkinje cell specific zebrin antigens in the cerebellum of the meander tail mutant mouse. Brain Res 1992; 589:135-138.

196. Napieralski JA, Eisenman LM. Developmental analysis of the external granular layer in the meander tail mutant mouse: Do cerebellar microneurons have independent progenitors? Dev Dynamics 1993; 197:244-254.

197. Bock GR, Frank MP. Brainstem responses in the quivering mutant mouse. Acta Otolaryngol 1984; 98:193-198.

198. Horner KC, Bock GR. Single unit responses in the cochlear nucleus of the deaf quivering mouse. Hearing Res 1984; 13:63-72.

199. Horner KC, Bock GR. Combined electrophysiological and autoradiographic delimitation of retrocochlear dysfunction in a mouse mutant. Brain Res 1985; 331:217-223.

200. Tong J, Potts JF, Rochelle JM et al. A single B-1 subunit mapped to mouse chromosome 7 may be a common component of Na channel isoforms from brain, skeletal muscle and heart. Biochem Biophys Res Commun 1993; 195:679-685.

201. Lane PW, Bronson RT, Spencer CA. Rostral cerebellar malformation, *(rcm)*: A new recessive mutation on chromosome 3 of the mouse. J Hered 1992; 83:315-318.

202. Caddy KWT, Sidman RL. Purkinje cells and granule cells in the cerebellum of the stumbler mutant mouse. Dev Brain Res 1981; 1:221-236.

203. Caddy KWT, Patterson DL, Biscoe TJ. Use of the UCHT1 monoclonal antibody to explore mouse mutants and development. Nature (Lond) 1982; 300:441-443.

204. Turgeon SM, Albin RL. Pharmacology, distribution, cellular localization, and development of $GABA_B$ binding in rodent cerebellum. Neuroscience 1993; 55:311-323.

205. Thomas KR, Musci TS, Neumann PE et al. Swaying is a mutant allele of the proto-oncogene *Wnt-1*. Cell 1991; 67:969-976.

206. Rossi F, Jankovski A, Sotelo C. Target neuron controls the integrity of afferent axon phenotype: A study on the Purkinje cell-climbing fiber system in cerebellar mutant mice. J Neurosci 1995; 15:2040-2056.

207. Feddersen RM, Ehlenfeldt R, Yunis WS et al. Disrupted cerebellar cortical development and progressive degeneration of Purkinje cells in SV40 T antigen transgenic mice. Neuron 1992; 9:955-966.

208. Luo L, Hensch TK, Ackerman L et al. Differential effects of the Rac GTPase on Purkinje cell axons and dendritic trunks and spines. Nature (Lond) 1996; 379:837-840.

209. Koike T, Tanaka S, Ito E. Neuronal development and apoptosis. Human Cell 1994; 7:13-19.

CHAPTER 5

Neurological Mutant Mice as Genetic Models for Neuronal Transplantation

INTRODUCTION

There is a large number of mutations in the laboratory mouse that affect the development or maintenance of specific cell populations or regions of the nervous system.[1,2] Such mutations provide useful experimental material for the study of cellular phenomena related to normal and aberrant development of the brain and the degeneration of nerve cells.

Some of the molecular mechanisms by which neurological mutations lead to the expression of the corresponding phenotypes have begun to be undeciphered (cf. chapter 4). In specific cases, experimental work with mouse chimeras has provided suggestive evidence for the intrinsic action of the mutant allele within the affected neuronal population.[3] In other cases, the neuronal or systemic environment may play a part in the pathogenesis of the condition.

The expansion of the techniques of neural transplantation[4] have made it possible to employ neurological mutant mice in such research. Mouse mutants can be used both *(i)* to obtain donor tissue, in order to follow growth properties in a different cellular environment and thus learn about possible cellular mechanisms leading to degeneration and *(ii)* as recipients of the appropriate, genetically-normal tissue, in attempts to restore anatomical and functional deficits resulting from the genetic mutation.

In interpreting results from such experimentation, one ought to keep in mind that in addition to features of transplant growth inherent in the process of removing a piece of tissue from its own location and leading it to develop inside a new milieu, there are potential cellular interactions between donor and host tissues, both in mutant-to-normal and normal-to-mutant grafting experiments, that might involve: *(i)* humoral interactions at the systemic level and *(ii)* neuronal or glial interactions at the organ or micro-environmental level. Specific examples are presented later on in this chapter.

When it comes to the use of mutant mice as models of neurological disorders, there is always the advantage of having a lesion that is relatively consistent and does not require induction by any treatment, and of a natural disease that may be anatomically comparable to human neurological conditions. In the latter context in particular, mutant mice can be useful models for experimentation pertinent to the application of neural transplantation strategies in combating progressive neurological disorder.

Alternative approaches for the correction of neurological deficits in mutant mice include the use of genetic engineering techniques such as the production of transgenic organisms.[5-7] Neural transplantation and gene injection experimentation are not mutually exclusive, but rather, complementary. Transgenic technology is applicable at much earlier stages of embryo growth and is conceptually curative. On the other hand, neural transplantation may constitute a symptomatic treatment, but it can be applied practically at any stage of the life of the recipient organism. In either case, however, the correction of a deficit has been only partial so far.

In this chapter, an overview is given of neurological mutants that have been used in neuronal transplantation research (Table 5.1). Transplantation of oligodendroglia and peripheral nerve has also been carried out in myelin mutants or mutants with peripheral neuropathies;[8-12] however, the present chapter is confined to studies dealing with nerve cell transplants. A brief overview of the neurological deficit is followed by an account of the

Table 5.1. Diagrammatic summary of neurological mutant mice used as models for neuronal transplantation experiments (excluding cerebellar models)

Mutant mouse	Gene symbol	Major deficit	Donor tissue	Host	Refs.
Eyeless	ey	Anophthalmia	Mutant eye anlage	Normal mouse embryos	15
Ocular retardation	or	Anophthalmia	Fetal neural retina	Mutant mice	16
Retinal degeneration	rd	Outer segment and rod cell degeneration	Normal neonatal mouse retina	Mutant mice	25-27
Hypogonadal	hpg	GnRH deficiency	Normal fetal preoptic area	Mutant mice	30-47
Reeler	rl	Abnormal organization of cortex, cerebellum and hippocampus	Neonatal mutant or normal hippocampus	Normal or mutant mice	53
Weaver	wv	Loss of mesencephalic dopamine neurons	Normal or heterozygous fetal midbrain	Mutant mice	57,60, 63-72

questions asked and of the results obtained from transplantation experiments in each model.

VISUAL SYSTEM MODELS

EYELESS MUTANT

The eyeless *(ey)* mutation, described in the ZRDCT strain by Chase and Chase,[13] is recessive and responsible for complete anophthalmia in 70% of the adults and for unilateral or bilateral microphthalmia in the remaining 30%.[14] In homozygous mutants *(ey/ey)*, the eye primordium appears and begins to differentiate on embryonic day 9 (E9); on E10, lens invagination is abnormal, the

lens is smaller than normal and improperly centered in the optic
cup and by E13, the primordial eye has regressed completely.

Salaün[15] grafted mutant eye anlagen intrafetally into normal
hosts to determine whether by being withdrawn from the environment of the mutant embryo and being placed into a normal fetal
environment, the expression of the mutation could be modified
and the eye rescued. Mutant donor tissue was taken at E10; normal host fetuses were E14-E17. Grafts were inserted under the
skin of the back of the recipient embryos in areas close to the
head. The results showed that one week after transplantation grafts
were recovered with a well-differentiated neural retina in 60-64%
of the cases; in one-third of the cases, a lens appeared as well.
These findings suggest that when withdrawn from its specific embryonic environment, the *ey* ophthalmic anlage is able to differentiate much better than in situ and that a fairly typical differentiation of the retina can be obtained. Therefore, it seems likely that
eye regression in the mutant is not the expression of an intrinsic
deficit in the organ, but it may rather result from deleterious
influences of the mutant environment, local or systemic.

OCULAR RETARDATION MUTANT

Hankin and Lund[16] used another strain of congenitally
anophthalmic mice, the "ocular retardation" mutant (129/SV-CP
or ʲ /*or* ʲ), as recipients of E13 normal CD-1 neural retinal transplants. The eyecup of homozygous "ocular retardation" mice de-
velops abnormally, and the optic nerve is aplastic, leading to eye
degeneration and effective anophthalmia. Hosts were two to three
days old at the time of transplantation and were allowed survival
times of three weeks postoperatively. The idea of the study was to
test whether prior optic innervation of the superior colliculus plays
a role in establishing either of two components of optic axon outgrowth in the retinotectal system: elongation along a "subpial pathway" in the rostral brainstem or "target-directed outgrowth" along
the midbrain parenchyma. The results showed that the former type
of innervation from grafts placed on the surface of the brainstem
may take place in the absence of prior innervation of the superior
colliculus; the latter form of axonal outgrowth from grafts embed-
ded into the midbrain parenchyma occurs only if the tectum is

also innervated by a second graft placed on the surface of the brainstem. Thus, it appears that the guidance of optic axons to the superior colliculus is a complex process that may depend on elaborate interactions between polarized substrate-dependent growth and one or more target-derived cues.

RETINAL DEGENERATION MUTANT

The retinal degeneration *(rd)* mutation was described by Brückner[17] and Tansley[18] and is recessive. In homozygous mutant mice *(rd/rd)*, eyes develop normally until postnatal day 7-10; then the outer segments and the rod cells degenerate, such that by day 15 only a thin layer of rod cells is left, and by day 35 they have disappeared completely.[19,20] Although the inner nuclear layer and the retinal ganglion cells appear normal, a slight quantitative reduction has been described.[21,22] Cones degenerate at a much slower rate than rods, such that a few are still present at 18 months of age.[23] Experiments with mouse chimeras indicate that the *rd* locus does not act in the pigment epithelium, but probably in the photoreceptor cells.[24]

Fragments of neonatal mouse retina were transplanted into the posterior pole of the eye of adult *rd/rd* mutant hosts.[25,26] By 15 days after transplantation, grafted tissue grew into the retina and subretinal space of the host; differentiated outer nuclear layer photoreceptors were seen from which the hosts had been previously deprived. Furthermore, transplanted cells expressed immunoreactivity for S-antigen, a specific marker for photoreceptor cells. These results suggested that developing retina can be transplanted successfully into an extensively damaged host eye; such a procedure may replace missing neuroretinal cells.

In a study by a different group,[27] microaggregates of one-to two-day-old neural retina were transplanted into the subretinal space of adult *rd* mutant and normal mice, and survival times of up to nine months were allowed postoperatively. The results showed that donor cones and rods develop and survive well if grafted with the proper orientation to the retinal pigment epithelium of *rd* mutant hosts. Grafts to normal hosts also develop normally and displace host photoreceptors from the pigment epithelium, which eventually degenerate. A barrier is formed by Müller cell processes, which demarcates host from transplanted tissue. In all, those data confirm the notion that progenitor photoreceptors transplanted into the mutant eye survive well and reconstitute a photoreceptor layer.

A six-week delay in photoreceptor cell death has been achieved in *rd* mutant mice after subretinal injection of a recombinant replication-defective adenovirus containing the murine cDNA for the wild-type β subunit of the cGMP phosphodiesterase gene.[28]

NEUROENDOCRINE SYSTEM

The hypogonadal mutation *(hpg)* was found on a C3H/HeH × 101/H background and is recessive. Both male and female homozygous mice *(hpg/hpg)* have underdeveloped reproductive tracts and are infertile, due to a deficiency of hypothalamic gonadotrophin-releasing hormone (GnRH), resulting in a reduction in pituitary content and circulating levels of luteinizing hormone (LH) and follicle-stimulating hormone (FSH).[29] In homozygous males the testes are small and located in the abdomen, and spermatogenesis is arrested; in homozygous females the ovaries and uterus are underdeveloped.

Grafts of normal fetal preoptic area tissue have been implanted into the third ventricle of adult hypogonadal mutant mice; GnRH-containing neurons were detected in the grafts by immunocytochemistry; following transplantation, mating, pregnancy and delivery of healthy litters were achieved.[30-47]

HIPPOCAMPAL GRAFTS

The reeler mutation *(rl)* is recessive and occurred spontaneously in an inbred stock of mice.[48] Homozygous mutants *(rl/rl)* display an alteration of the size and typical organization and lamination of the cerebellar cortex, the cerebral cortex and hippocampus.[49] Developmental autoradiography studies have shown that different classes of neurons originating in the ependymal layer at the normal time migrate abnormally, coming to rest in abnormal relations with each other.[50,51] In the hippocampus, in particular, pyramidal and granule cells are not arranged in tightly-packed layers, but are dispersed; furthermore, the innervation of hippocampal neurons by afferent fibers can be abnormal with respect to both type of cell and cellular site of innervation.[52] Behaviorally, homozygous reeler mice are unable to keep their hindquarters upright and frequently fall over on their sides when walking or running.[1]

In order to establish whether the aberrant connectivity seen in the reeler mutant reflects an abnormality of the intrinsic hippocampal cells or of their environment, pieces of hippocampal tissue were transplanted from neonatal mutant mouse into neonatal normal mouse and from neonatal normal mouse into neonatal mutant mouse.[53] Mutant donor transplanted into normal host showed a few pyramidal cells and many granule cells, without, however, a lamination of the latter. Normal donor grafted into a mutant host consisted of granule cells curled into a layer, with a few scattered pyramidal neurons. It was concluded that granule cells from normal tissue retain their laminated organization in the reeler environment, whereas the reeler hippocampus retains its phenotypic organization when transplanted into the normal brain.

MESOTELENCEPHALIC DOPAMINE PROJECTION SYSTEM

A genetic deficit of the mesotelencephalic dopamine (DA) projection system has been identified in weaver mutant mice;[54] the striatal DA deficiency is 75% relative to controls and is accompanied by a 27% DA deficiency in the olfactory tubercle and a 77% DA deficiency in frontal cortex.[55] At the time when striatal DA concentration in weaver mutants is at its peak level, i.e., on postnatal day 20, the overall incidence of junctional synapses formed by DA afferents in the weaver striatum is decreased relative to normal, thus adding further to the functional deficit.[56]

Neuronal loss occurs in all groups of the mesotelencephalic DA cell system.[57,58] By three months of age, the DA cell deficit in homozygous weaver mutants amounts to 56% in retrorubral nucleus (area A8), 69% in substantia nigra (area A9) and 26% in ventral tegmental area (area A10). The losses of DA neuron somata account for DA deficiency in telencephalic terminal fields normally receiving DA innervation from those cells.

The value of using mutant animal models of neurodegenerative diseases in the application of neural transplantation has been emphasized; in particular, the weaver mouse has been established as the only available model of naturally-occurring degeneration in the nigrostriatal dopamine system suitable for neural transplantation studies pertaining to Parkinson's disease.[59-62]

To examine the ability of genetically-normal DA-containing grafts to compensate for the genetic mesostriatal deficit of weaver homozygotes, ventral mesencephalic anlagen, containing the primordial substantia nigra and prepared from wild-type embryos, were implanted either as solid grafts into the lateral ventricle[57] or into a cortical cavity over the dorsal neostriatum[63-65] of adult weaver mutants or as cell suspensions injected stereotactically into the striatal parenchyma.[66-72] Survival times of one to nine months after grafting have been allowed.

Wild-type grafts survive in the mutant environment and contain about 100-1000 tyrosine hydroxylase (TH) immunoreactive neurons. TH immunoreactive fibers, displaying their typical varicosities, reach a depth of innervation that covers the dorsal one-half of the striatal aspect at 4.5 months after grafting.

Unilateral transplantation of ventral mesencephalic grafts leads to a circling preference toward the nongrafted side in rotational asymmetry tests after methamphetamine administration, indicating that grafted DA neurons establish a functional innervation of the weaver striatum.[64,66,67] Bilateral grafts of mesencephalic grafts lead to improvement in a battery of behavioral tests, including balance rod, open-field and locomotor tasks.[68]

Ultrastructural studies on the synaptic relationships of TH immunoreactive axon terminals in the grafted weaver striatum have disclosed percentages that approximate the values found in normal animals.[63,65,67] The majority of contacts in the reinnervated striatum (84%) are made with dendritic spines and shafts. The occurrence of newly formed synapses in the mutant striatum after transplantation indicates that grafted DA neurons may exert their functional effects, at least in part, through synaptic activity.

In addition to axons, grafts also provide a dendritic innervation to the weaver striatum.[67] However, it appears that the axonal innervation is mostly responsible for the induction of rotational asymmetry in the unilateral transplantation setting.[69]

Mesencephalic cell suspension grafts in the weaver neostriatum restore parameters of DA uptake[66,70] and express many normal structural molecules.[71,72] Despite all of these effects, there is indicative evidence that donor DA neurons survive in smaller numbers when grafted to the striatum of weaver mutant hosts than when grafted into the striatum of wild-type mice subjected to 6-hydroxydopamine lesions of the substantia nigra.[69]

CEREBELLAR MODELS

The structural integration of cerebellar grafts into the cerebellum of mutant mice and their functional effects are discussed in chapters 7 and 8.

REFERENCES

1. Green MC, ed. Genetic Variants and Strains of the Laboratory Mouse. Stuttgart, New York: Gustav Fischer Verlag, 1981.
2. Lyon MF, Searle AG, eds. Genetic Variants and Strains of the Laboratory Mouse. 2nd ed. Oxford: Oxford University Press, 1989.
3. Mullen RJ, Herrup K. Chimeric analysis of mouse cerebellar mutants. In: Breakfield XO, ed. Neurogenetics: Genetic Approaches to the Nervous System. Amsterdam: Elsevier/North Holland, 1979:173-196.
4. Björklund A, Brundin P, Isacson O. Neuronal replacement by intracerebral neural implants in animal models of neurodegenerative disease. Adv Neurol 1988; 47:455-492.
5. Mason AJ, Pitts SL, Nikolics K et al. The hypogonadal mouse: Reproductive functions restored by gene therapy. Science 1986; 234:1372-1378.
6. Readhead C, Popko B, Takahashi N et al. Expression of a myelin basic protein gene in transgenic shiverer mice: Correction of the dysmyelinating phenotype. Cell 1987; 48:703-712.
7. Popko B, Puckett C, Lai E et al. Myelin deficient mice: Expression of myelin basic protein and generation of mice with varying levels of myelin. Cell 1987; 48:713-721.
8. Aguayo AJ, Attiwell M, Trecarten J et al. Abnormal myelination in transplanted trembler mouse Schwann cells. Nature (Lond) 1977; 265:73-75.
9. Gansmüller A, Lachapelle F, Baron Van Evercooren A et al. Transplantations of newborn CNS fragments into the brain of shiverer mutant mice: Extensive myelination by transplanted oligodendrocytes. II. Electron microscopic study. Dev Neurosci 1986; 8:197-207.
10. Kohsaka S, Yoshida K, Inoue Y et al. Transplantation of bulk-separated oligodendrocytes into the brains of shiverer mutant mice: immunohistochemical and electron microscopic studies on the myelination. Brain Res 1986; 372:137-142.

11. Friedman E, Nilaver G, Carmel P et al. Myelination by transplanted fetal and neonatal oligodendrocytes in a dysmyelinating mutant. Brain Res 1986; 378:142-146.

12. Lubetzki C, Gansmüller A, Lachapelle F et al. Myelination by oligodendrocytes isolated from 4-6-week-old rat central nervous system and transplanted into newborn shiverer brain. J Neurol Sci 1988; 88:161-175.

13. Chase HB, Chase EB. Studies on an anophthalmic strain of mice. I. Embryology of the eye region. J Morphol 1941; 68:279-301.

14. Beck SL. The anophthalmic mutant of the mouse. I. Genetic contribution of the anophthalmic phenotype. J Hered 1963; 54:39-44.

15. Salaün J. Differentiation of the optic cups from an anophthalmic murine strain, in culture and in intrafoetal grafts. J Embryol Exp Morphol 1982; 67:71-80.

16. Hankin MH, Lund RD. Induction of target-directed optic axon outgrowth: Effect of retinae transplanted to anophthalmic mice. Dev Biol 1990; 138:136-146.

17. Brückner R. Spaltlampenmikroskopie und Ophthalmoskopie am Auge von Ratte und Maus. Doc Ophthalmol 1951; 5-6:452-554.

18. Tansley K. An inherited retinal degeneration in the mouse. J Hered 1954; 45:123-127.

19. Caley DW, Johnson C, Liebelt RA. The postnatal development of the retina in the normal and rodless CBA mouse: A light and electron microscopic study. Am J Anat 1972; 133:179-212.

20. LaVail MM, Sidman RL. C57BL/6J mice with inherited retinal degeneration. Arch Ophthalmol 1974; 91:394-400.

21. Blanks JC, Bok D. An autoradiographic analysis of postnatal cell proliferation in the normal and degenerative mouse retina. J Comp Neurol 1977; 174:317-328.

22. Grafstein B, Murray M, Ingoglia NA. Protein synthesis and axonal transport in retinal ganglion cells of mice lacking visual receptors. Brain Res 1972; 44:37-48.

23. Carter-Dawson LD, LaVail MM, Sidman RL. Differential effect of the *rd* mutation on rods and cones in the mouse retina. Invest Ophthalmol Vis Sci 1978; 17:489-498.

24. LaVail MM, Mullen RJ. Role of the pigment epithelium in inherited retinal degeneration analyzed with experimental mouse chimeras. Exp Eye Res 1976; 23:227-245.

25. del Cerro M. Retinal transplants. Prog Retinal Res 1990; 9:229-272.

26. Jiang L-Q, del Cerro M. Reciprocal retinal transplantation: A tool for the study of an inherited retinal degeneration. Exp Neurol 1992; 115:325-334.

27. Gouras P, Du J, Kjeldbye H et al. Long-term photoreceptor transplants in dystrophic and normal mouse retina. Invest Ophthalmol Vis Sci 1994; 35:3145-3153.

28. Bennett J, Tanabe T, Sun D et al. Photoreceptor cell rescue in retinal degeneration *(rd)* mice by in vivo gene therapy. Nature Med 1996; 2:649-654.

29. Cattanach BM, Iddon CA, Charlton HM et al. Gonadotropin-releasing hormone deficiency in a mutant mouse with hypogonadism. Nature (Lond) 1977; 269:338-340.

30. Krieger DT, Perlow MJ, Gibson MJ et al. Brain grafts reverse hypogonadism of gonadotropin-releasing hormone deficiency. Nature (Lond) 1982; 298:1-3.

31. Gibson MJ, Krieger DT, Charlton HM et al. Mating and pregnancy can occur in genetically hypogonadal mice with preoptic area brain grafts. Science 1984; 225:949-951.

32. Silverman A-J, Zimmerman EA, Gibson MJ et al. Implantation of normal fetal preoptic area into hypogonadal *(hpg)* mutant mice: Temporal relationships of the growth of GnRH neurons and the development of the pituitary/testicular axis. Neuroscience 1985; 16:69-84.

33. Gibson MJ, Krieger DT. Neuroendocrine brain grafts in mutant mice. Trends Neurosci 1985; 8:331-334.

34. Silverman A-J, Zimmerman EA, Kokoris GJ et al. Ultrastructure of gonadotropin-releasing hormone neuronal structures derived from normal fetal preoptic area and transplanted into hypogonadal *(hpg)* mice. J Neurosci 1986; 6:2090-2096.

35. Gibson MJ, Moscovitz HC, Kokoris GJ et al. Plasma LH rises rapidly following mating in hypogonadal female mice with preoptic area (POA) brain grafts. Brain Res 1987; 424:133-138.

36. Silverman A-J, Kokoris GJ, Gibson MJ. Quantitative analysis of synaptic input to gonadotropin-releasing hormone neurons in normal mice and *hpg* mice with preoptic area grafts. Brain Res 1988; 443:367-372.

37. Kokoris GJ, Lam NY, Ferin M et al. Transplanted gonadotropin-releasing hormone neurons promote pulsatile luteinizing hormone secretion in congenitally hypogonadal *(hpg)* male mice. Neuroendocrinology 1988; 48:45-52.

38. Silverman RC, Silverman A-J, Gibson MJ. Identification of gonadotropin releasing hormone (GnRH) neurons projecting to the median eminence from third ventricular preoptic area grafts in hypogonadal mice. Brain Res 1989; 501:260-268.

39. Broadwell RD, Charlton HM, Ganong WF et al. Allografts of CNS tissue possess a blood-brain barrier. I. Grafts of medial preoptic area in hypogonadal mice. Exp Neurol 1989; 105:135-151.

40. Gibson MJ, Silverman RC, Silverman A-J. Current progress in studies of GnRH cell-containing brain grafts in hypogonadal mice. Prog Brain Res 1990; 82:169-178.

41. Silverman RC, Gibson MJ, Silverman A-J. Relationship of glia to GnRH axonal outgrowth from third ventricular grafts in *hpg* hosts. Exp Neurol 1991; 114:259-274.

42. Livne I, Gibson MJ, Silverman A-J. Brain grafts of migratory GnRH cells induce gonadal recovery in hypogonadal *(hpg)* mice. Dev Brain Res 1992; 69:117-123.

43. Silverman A-J, Roberts JL, Dong KW et al. Intrahypothalamic injection of a cell line secreting gonadotropin-releasing hormone results in cellular differentiation and reversal of hypogonadism in mutant mice. Proc Natl Acad Sci USA 1992; 89:10668-10672.

44. Gibson MJ, Friedrich VL, Elder G et al. Human midsized neurofilament expression in transgenic mouse-derived grafts facilitates study of graft-host interactions in hypogonadal mice. Cell Transpl 1993; 2:223-227.

45. Miller GM, Silverman A-J, Roberts JL et al. Functional assessment of intrahypothalamic implants of immortalized gonadotropin-releasing hormone-secreting cells in female hypogonadal mice. Cell Transpl 1993; 2:251-257.

46. Gibson MJ, Silverman A-J. Neuroendocrine brain grafts. Semin Neurosci 1993; 5:423-430.

47. Miller GM, Silverman A-J, Rogers MC et al. Neuromodulation of transplanted gonadotropin-releasing hormone neurons in male and female hypogonadal mice with preoptic area brain grafts. Biol Reprod 1995; 52:572-583.

48. Falconer DS. Two new mutants, "trembler" and "reeler", with neurological actions in the house mouse. J Genet 1951; 50:192-201.

49. Sidman RL. Development of interneuronal connections in brains of mutant mice. In: Carlson FD, ed. Physiological and Biochemical Aspects of Nervous Integration. Englewood Cliffs, NJ: Prentice-Hall, 1968:163-193.

50. Caviness VS. Time of neuron origin in the hippocampus and dentate gyrus of normal and reeler mutant mice: An autoradiographic analysis. J Comp Neurol 1973; 151:113-120.

51. Caviness VS, Sidman RL. Retrohippocampal, hippocampal and related structures of the forebrain in the reeler mutant mouse. J Comp Neurol 1973; 147:235-254.

52. Stanfield BB, Caviness VS, Cowan WM. The organization of certain afferents to the hippocampus and dentate gyrus in normal and reeler mice. J Comp Neurol 1979; 185:461-484.

53. Errington ML, Bliss TVP. Hippocampal transplants in normal and reeler mice. Neurosci Lett 1984; 45:291-296.

54. Lane JD, Nadi NS, McBride WJ et al. Contents of serotonin, nore-pinephrine and dopamine in the cerebrum of the 'staggerer', 'weaver' and 'nervous' neurologically mutant mice. J Neurochem 1977; 29:349-350.

55. Schmidt MJ, Sawyer BD, Perry KW et al. Dopamine deficiency in the weaver mutant mouse. J Neurosci 1982; 2:376-380.

56. Triarhou LC, Norton J, Ghetti B. Synaptic connectivity of tyrosine hydroxylase immunoreactive nerve terminals in the striatum of normal, heterozygous and homozygous weaver mutant mice. J Neurocytol 1988; 17:221-232.

57. Triarhou LC, Low WC, Ghetti B. Transplantation of ventral mesencephalic anlagen to hosts with genetic nigrostriatal dopamine deficiency. Proc Natl Acad Sci USA 1986; 83:8789-8793.

58. Triarhou LC, Norton J, Ghetti B. Mesencephalic dopamine cell deficit involves areas A8, A9 and A10 in weaver mutant mice. Exp Brain Res 1988; 70:256-265.

59. Sinden JD, Patel SN, Hodges H. Neural transplantation: Problems and prospects for therapeutic application. Curr Opin Neurol Neurosurg 1992; 5:902-908.

60. Triarhou LC, Low WC, Doucet G et al. The weaver mutant mouse as a model for intrastriatal grafting of fetal dopamine neurons. In: Hefti F, Weiner WJ, eds. Progress in Parkinson's Disease Research—2. Mount Kisco, NY: Futura Publishing, 1992:389-400.

61. Bankiewicz K, Mandel RJ, Sofroniew M. Trophism, transplantation, and animal models of Parkinson's disease. Exp Neurol 1993; 124:140-149.

62. Brundin P, Duan W-M, Sauer H. Functional effects of mesencephalic dopamine neurons and adrenal chromaffin cells grafted to the rodent striatum. In: Dunnett SB, Björklund A, eds. Functional Neural Transplantation. New York: Raven Press, 1994:9-46.

63. Triarhou LC, Low WC, Ghetti B. Synaptic investment of striatal cellular domains by grafted dopamine neurons in weaver mutant mice. Naturwissenschaften 1987; 74:591-593.

64. Low WC, Triarhou LC, Kaseda Y et al. Functional innervation of the striatum by ventral mesencephalic grafts in mice with inherited nigrostriatal dopamine deficiency. Brain Res 1987; 435:315-321.

65. Triarhou LC, Low WC, Norton J et al. Reinstatement of synaptic connectivity in the striatum of weaver mutant mice following transplantation of ventral mesencephalic anlagen. J Neurocytol 1988; 17:233-243.

66. Doucet G, Brundin P, Seth S et al. Degeneration and graft-induced restoration of dopamine innervation in the weaver mouse neostriatum: a quantitative radioautographic study of [^{3}H]dopamine uptake. Exp Brain Res 1989; 77:552-568.

67. Triarhou LC, Brundin P, Doucet G et al. Intrastriatal implants of mesencephalic cell suspensions in weaver mutant mice: ultrastructural relationships of dopaminergic dendrites and axons issued from the graft. Exp Brain Res 1990; 79:3-17.

68. Triarhou LC, Norton J, Hingtgen JN. Amelioration of the behavioral phenotype in weaver mutant mice through bilateral intrastriatal grafting of fetal dopamine cells. Exp Brain Res 1995; 104:191-198.

69. Witt TC, Triarhou LC. Transplantation of mesencephalic cell suspensions from wild-type and heterozygous weaver mice into the denervated striatum: Assessing the role of graft-derived dopaminergic dendrites in the recovery of function. Cell Transpl 1995; 4:323-333.

70. Triarhou LC, Stotz EH, Low WC et al. Studies on the striatal dopamine uptake system of weaver mutant mice and effects of ventral mesencephalic grafts. Neurochem Res 1994; 19:1349-1358.

71. Solà C, Mengod G, Low WC et al. Regional distribution of amyloid β-protein precursor, growth-associated phosphoprotein-43 and microtubule-associated protein 2 mRNAs in the nigrostriatal system of normal and weaver mutant mice and effects of ventral mesencephalic grafts. Eur J Neurosci 1993; 5:1442-1454.

72. Triarhou LC, Solà C, Mengod G et al. Ventral mesencephalic grafts in the neostriatum of the weaver mutant mouse: Structural molecule and receptor studies. Cell Transpl 1995; 4:39-48.

CHAPTER 6

BASIC STUDIES ON CEREBELLAR TISSUE TRANSPLANTATION

INTRODUCTION

As a rule, the genesis of neuronal populations, including Purkinje cells, is concluded during embryonic life, and the regenerative capacity of the adult CNS is confined to compensatory fiber sprouting and not mitotic divisions of nerve cells.[1] Therefore, neurons that die as a result of regressive processes can only be replaced by implantation after harvesting from an external source. Intracerebral grafting of developing neuroblasts into the adult pathological brain has been successfully used to replace degenerated neurons in several experimental instances.[2,3] In particular, primordial cerebellar tissue has been shown to survive and grow after orthotopic or heterotopic implantation into the adult rodent brain. An account of these studies is presented in this chapter. Cerebellar neuron grafting has also been applied to neurological mutant mice both to create appropriate confrontations between wild-type and mutant cells in elucidating gene effects on the involved lineage and to study the structural integration of transplanted wild-type Purkinje cells into the disrupted cerebellar loop; an account of cerebellar transplantation studies using ataxic mouse mutants is given in chapter 7.

In a sense, a cerebellar transplant in the brain or in the anterior eye chamber constitutes an "in vivo culture" that shares features with the growth properties of the cerebellum in tissue culture in vitro. The literature on the properties of cerebellum

in culture is large and will not be dealt with here. Instead, the
reader is referred to appropriate references.[4-6]

SURVIVAL OF CEREBELLAR PRIMORDIA, HISTOTYPIC DIFFERENTIATION AND SYNAPTOGENESIS OF CEREBELLAR GRAFTS

Das and Altman[7] transplanted slabs of infant (postnatal day 7,
P7) rat cerebellum into the cerebella of age-matched rat hosts, after labeling donor tissue with [³H]thymidine to tag cells undergo-
ing mitotic divisions. Analyzing a series of grafts that ranged in
survival times from 3 hours to 16 days, they found that transplanted, undifferentiated cells had migrated into the cerebellar cortex of the hosts, where they had differentiated into basket cells in
the molecular layer and into granule cells in the internal granule
cell layer. That study indicated that transplantation of neuronal
precursors is possible in the maturing nervous system of postnatal
mammals and attributed such success to two important factors,
the active migratory behavior of donor tissue and the ongoing gen-
erative process in the host cerebellum. In a more detailed account
that followed[8] pathological changes were also reported in grafts
with short-term survival, along with the aggregation of surviving
and proliferating cells into areas representing the viable portions
in the external germinal layer of grafted tissue. The issue of poor
viability of postmigratory elements such as Purkinje and Golgi cells
of the transplants was raised and the idea of using germinal cells
of the neuroepithelium at embryonic stages proposed.[8] Experiments
dealing with homotopic transplantation of cerebellar tissue in
neonatal rabbits were conducted as well.[9]

Hine[10] implanted embryonic day 18 (E18) rat cerebellar primordia into the rostral forebrain of P7 rat hosts and examined
histologically the growth and differentiation of the transplanted
tissue at various times after transplantation. The grafts grew and
acquired the trilaminar structure characteristic of the organotypic
differentiation of normal cerebellum. Nonspecific host afferents into
the transplant were found by using the Fink-Heimer method.

Wells and McAllister[11] transplanted E18 rat cerebellar primordia into the neocortex of P10-P12 rats and studied in great detail
the histological development of 53 transplants at survival times of

5 minutes to 426 days after grafting. They found that transplants grown on the neocortical surface had normal orientation and foliation patterns; on the other hand, pieces of cerebellar cortex confined within the depths of either the graft or the host tissue were layered normally, but the layers were arranged in concentric cylinders around blood vessels. Overall, the transplantation procedure did not delay the normal time sequence of prominent features of cerebellar development, such as initial molecular layer formation, peak development of the external germinal layer, completion of neuroblast migration from the external germinal layer and postmigratory changes within the transplants; however, subtle differences were noted in Purkinje cell differentiation, in particular involving processes of monolayer formation and foliation.

Alvarado-Mallart and Sotelo[12] transplanted pieces of E14-E15 rat cerebellum into a cortical cavity in the occipital lobe of 2-month-old rats. They allowed graft survival times of 2-3 months and analyzed the cytoarchitectonic and synaptic organization of 16 such transplants by light and electron microscopy. They described the growth and development of donor tissue into a cerebellar structure containing cortical and nuclear regions, with all five categories of neurons normally found in cerebellar cortex present in the former and with normal lamination and foliation pattern. Qualitatively normal synaptic connections were found among the various neuronal elements of the grafts with the exception of climbing fibers, which were obviously missing from the donor cerebellar tissue. Heterologous (atypical) synaptic arrangements were also encountered in the grafts, consisting of pseudoglomerular formations of tightly packed small axon terminals of unestablished origin with granule cell dendrites in the neuropil of the granule cell layer. Reciprocal connections were found between the cortical and nuclear regions of the transplants by means of horseradish peroxidase tracing experiments, providing further evidence for the organotypic, histotypic and synaptotypic differentiation of the cerebellar anlage after heterotopic transplantation into a cortical cavity in the adult rat brain.

Kromer et al[13,14] implanted E12-E13 and E17-E19 rat cerebellar primordia into cavities prepared ahead of time in the occipital-retrosplenial cortex or in the parietal cortex and septal pole of the hippocampus, and studied survival times of 6 weeks to 14 months after transplantation. They described subdivisions similar to deep cerebellar nuclei and a trilaminar cerebellar cortex in the heterotopic grafts, with development of cortical invagination in transplants of both stages. Nonetheless, early-gestation implants grew larger in size than late gestational tissues. In Golgi-impregnated specimens, Purkinje cells were found to possess well-developed dendritic arbors with smooth primary branches and studded-with-spines secondary and tertiary branches. The three-dimensional orientation of Purkinje dendritic trees was atypical, while the properties of the remaining four classes of cerebellocortical interneurons possessed a morphology very similar to normal adult cerebellum.

In a subsequent study, Ezerman and Kromer[15] prepared dissociated cell suspensions of E13 rat cerebellar primordia and reaggregated them into tissue pellets by centrifugation. After implantation into cortical cavities in adult rats and by analyzing survival times of 2, 4 and 6 weeks, they observed an initial sorting of macroneurons (i.e. Purkinje cells and deep nuclei neurons), followed by segregation of developing cortical cells into a trilaminar structure. Going a step further, Ezerman[16] described the survival and development of fetal and postnatal cerebellar grafts after growth in culture in the form of explants.

The organotypic and histotypic differentiation of cerebellar grafts transplanted into the lateral ventricle or hemispheric cerebral parenchyma of adult Wistar rats has been also described by Alexandrova and Polezhaev.[17] Kikuchi[18] transplanted E14-E20 rat cerebellar primordia into the cerebellum of adult Fischer 344 rats and described normal synaptic connections between neuronal elements in the graft by electron microscopy at survival times of one month to one year after transplantation.

Chopko and Voneida[19] placed grafts of E15 rat cerebellar tissue in a forebrain cortical cavity of 2.5- to 7.5-month-old rat hosts and studied four grafts at 12-151 days after transplantation. Viable grafts were identified with mature cellular elements, but they lacked the overall anatomical arrangement or characteristic cytoarchitecture of normal cerebellum. The less-than-optimum graft organization in that study could be related in part to the small sample size, as well as the single-step implantation procedure adopted as opposed to the delayed cavity transplantation proto-col[20] that offers far superior conditions for graft survival and growth.

The morphological maturation of cerebellar grafts has also been studied after homologous transplantation into the anterior eye chamber in the form of either single grafts[21,22] or of cerebellar-cerebrocortical cografts.[23,24] Observations in Golgi-Cox and tolui-dine blue histological preparations have indicated that grafts of E15-E16 rat cerebellar tissue in oculo grow into structures with a trilaminar organization that contain all types of cerebellar neurons.[22] Differentiated cerebellar glomeruli are found by electron micros-copy, containing mossy terminals that most likely originate in the cerebellar nuclei portion of the graft.[21] When cerebellar grafts are cocultivated with cerebral cortical grafts in the anterior eye cham-ber, both γ-aminobutyric acid (GABA)-immunoreactive and GABA-immunonegative mossy-like terminals, originating either in the cerebral cortical tissue or in the cerebellar nucleus in the absence of "natural" mossy fibers, are seen forming asymmetric synapses with granule cell dendritic digits within the graft.[23] Further, in the absence of "natural" climbing fibers, Purkinje dendritic shafts receive symmetrical synapses from GABA-immunopositive "foreign" climbing-like terminals, a phenomenon pointing to the plasticity of Purkinje cell dendrites in the absence of specific afferents.[24]

As retrovirus-mediated oncogene transfer technology became available, Wiestler et al[25,26] and Snyder et al[27] have used it in com-bination with the neural grafting model to study issues of the com-mitment and differentiation of cerebellar progenitor cells. Wiestler et al[25] introduced the v-Ha-*ras* and v-*myc* oncogenes into the de-veloping rat brain; introduction of the same oncogenes into new-born cerebellar cultures was affected as well. Their data showed a powerful complementary transforming effect of the two oncogenes on neural progenitors both in vivo and in vitro, suggesting that coexpression of the *ras* and *myc* oncogenes may provide a highly efficient tool for transforming neural precursor cells in distinct segments of CNS at different stages of development.

Snyder et al[27] generated multipotent neural cell lines via v-*myc* transfer into mouse cerebellar progenitor cells and transplanted them back into the cerebella of newborn mice. Transformed cells became integrated into the host cerebellum in a nontumorigenic and cytoarchitectonically appropriate manner, differentiating into neurons or glia depending on engraftment site. That study lent support to the idea that immortalized cell lines can be generated to repair or to deliver exogenous genes into the CNS.

Finally, retrovirus-mediated gene transfer has been used to transfect mouse cerebellar primary cultures with recombinant retroviruses harboring the bacterial enzyme chloramphenicol acetyltransferase prior to transplantation into adult mouse cerebellum.[28,29] Immunocytochemical analyses of the tissues with various antibodies indicated stable marking of labeled cells both in the grafts and in the host molecular layer.

BLOOD-BRAIN BARRIER OF CEREBELLAR GRAFTS

Heterotopic cerebellar grafts inserted stereotactically into the rat corpus striatum receive sufficient vascularization[30] and lend themselves to a variety of possible ways for in vivo manipulation, e.g., through induction of toxic metabolic states,[31,32] owing to their standardized position relative to bregma and dura mater. All ultrastructural elements of a normal blood-brain barrier are seen in the capillary vessels formed within such intrastriatal cerebellar grafts.[33]

The functional permeability to macromolecules in cerebellar grafts has been studied in solid grafts of fetal rat tissue implanted into the cerebral ventricles of young adult hosts, analyzed at 2-600 days after transplantation.[34] Fenestrated vessels were not directly observed, even though vessels indigenous to the grafts retained blood-brain barrier properties. Horseradish peroxidase (HRP), HRP-human serum albumin and HRP-human IgG given intravascularly 3-60 minutes before sacrifice showed that younger grafts were filled with the macromolecules, whereas older grafts displayed variability in permeation. HRP injections into the cerebrospinal fluid suggested that solutes may flow at an increased rate (up to 3-fold greater than normal) through the grafts.[34]

MITOTIC ACTIVITY

Purkinje cell progenitors in the cerebellar grafts proceed normally to conclude the final phase of neuroblastic proliferation in accordance with an undisturbed temporal window of mitotic activity, as determined both by [³H]thymidine autoradiography of solid E12 mouse cerebellar primordia transplanted into *pcd* mutant mice[35-37] and by 5'-bromodeoxyuridine (BrdU) labeling of solid E14 rat cerebellar primordia transplanted into the cerebellum of adult rats.[38,39]

In a cytophotometric study, E17 cerebellar grafts were implanted into the sensorimotor cortex of syngeneic rats, and the DNA content of donor Purkinje cells and granule cells was measured 30 days after transplantation;[40] it was found that about 3% of the transplanted Purkinje cells contained hyperdiploid and tetraploid nuclei, which corresponds to the percentage encountered in adult normal cerebellum. On the other hand, granule cells were diploid, which is the case normally as well.[40]

HISTOCHEMICAL PHENOTYPY OF TRANSPLANTED PURKINJE CELLS

Histochemical studies have shown transplanted Purkinje cells to express many of the structural, neurotransmitter-related and growth-factor system molecules that normal Purkinje cells contain in the intact cerebellum. In particular, Purkinje cells in cerebellar grafts selectively express 28 kDa Ca^{2+}-binding protein (CaBP or calbindin[41]),[42,43] guanosine 3',5'-phosphate-dependent protein kinase (cGK)[44] and polypeptide PEP-19.[45] Further, they express zebrin I in a topographic order that consists of immunopositively-defined compartments of longitudinal bands in grafts placed into the cerebellum of rats that had previously been subjected to kainic acid lesions[46] and in intraocular or intracortical-cavity grafts.[47] Transplanted Purkinje cells also immunoreact with monoclonal antibodies mabQ113[48] and mab-1D10,[49] anti-spot 35 antibody[38,39] and anti-Leu-4 (CD3).[33] Finally, Purkinje cells in cerebellar grafts express positive immunoreactivity for neurofilament protein, synapsin[50] and nonphosphorylated neurofilament epitopes (nPNF).[51]

With reference to neurotransmission-related molecules, transplanted Purkinje cells show motilin immunoreactivity.[52] Using a rabbit antiserum against synthetic peptides corresponding to sequences specific for the C-terminus of subunits 2 and 3 of the α-amino-3-hydroxy-5-methyl-4-isoxazole propionic acid (AMPA) class of glutamate receptors (anti-GluR2/3[53]), positive immunoreactivity was found in transplanted Purkinje cells[54,55] with the following dichotomy: neurons occupying cerebellocortical localities display GluR2/3 immunoreactivity both in their somata and dendritic trees, whereas neurons arrested intraparenchymally express GluR2/3 immunoreactivity only in the perikaryon, suggesting a possible regulation by transacting elements from host parallel fibers. In a similar context, quantitative autoradiographic studies with [^{3}H]6-cyano-7-nitro-quinoxaline-2,3-dione (CNQX) as a ligand for non-N-methyl-D-aspartate (non-NMDA) glutamate receptors have shown that grafts migrating to the cerebellar cortex of the host express a higher density of [^{3}H]CNQX binding sites than grafts arrested intraparenchymally.[56]

Regarding growth factor systems, transplanted Purkinje cells express brain-derived neurotrophic factor (BDNF) mRNA,[39] nerve-growth factor (NGF) receptor peptide,[39] insulin-like growth factor (IGF)-I mRNA and peptide, and type I IGF receptor mRNA.[57]

PHYSIOLOGICAL ACTIVITY
OF TRANSPLANTED PURKINJE CELLS

The functional maturation of cerebellar grafts has been studied electrophysiologically in solid grafts of fetal rat cerebellum placed either into the anterior chamber of the rat eye[58] or into the cerebella of P5-P7 and P13-P14 rat pups.[59,60] In those studies, the spontaneous discharge rate of transplanted Purkinje cells was slightly slower than normal, a fact ascribable at least in part to the absence from the grafts of high-frequency bursts normally caused by the excitatory input from climbing fibers. On the other hand, the fact that local stimulation of the graft surface causes both decreased and increased Purkinje cell discharges suggests a normally functioning neurotransmission from the inhibitory molecular interneurons and the excitatory granule cells to Purkinje cell dendrites, physiological characteristics quite similar to the normal cerebellum.[59,60] In a different study, fetal cerebellar tissue from E20-E25 rabbit brain implanted intraocularly into the anterior eye chamber of athymic nude rats grows, and at 15 weeks after transplantation in vivo recording of single neuron activity reveals normal discharge rates of neurons.[50]

MIGRATORY PHENOMENA

Purkinje cells from heterotopic rat cerebellar grafts migrate into the host brain over considerable distances into regions adjacent to the transplants.[52] Neuronal migration and cerebellar lamination in rat fetal cerebellar grafts placed into the cerebral ventricles, lateral hypothalamus or parietal cortex of adult rats are more frequent at earlier gestational periods among E16, E18, E20 and E22 donor tissues.[52] Kawamura et al[61] have shown that both granule and Purkinje cells can migrate into the mature cerebellar cortex of normal adult rats. Solid grafts of fetal rat cerebellum transplanted into the fourth ventricle of normal rats develop into minicerebella that grow either toward the dorsal surface of the brainstem or to the overlying cerebellar cortex.[62] Grafted Purkinje cells may migrate out of the solid grafts to a certain extent into the normal cerebellar cortex of the host.

A site that favors Purkinje cell survival and growth appears to be the dorsal cochlear nucleus,[63] a structure with structural homology to the cerebellum.[64] By grafting solid fetal cerebellar tissue into the fourth ventricle in apposition to the dorsal cochlear nucleus, Rossi and Borsello[63] found that large numbers of donor Purkinje cells migrate and develop in its superficial layers, passing through the various phases that characterize normal ontogeny. A chick/quail chimeric model with partial cerebellar grafts has been employed to analyze the origin and migration of cerebellar cells.[65]

GLIAL ISSUES

The astrocytic populations of cerebellar grafts have been studied using immunocytochemistry for glial fibrillary acidic protein (GFAP) and vimentin.[59,60] For those experiments, grafts of E13-E15 rat cerebellum were placed into the cerebellum of 1- to 2-week-old rat pups. GFAP immunoreactivity was found at a glial interface along the graft-host border, as well as in Bergmann glial fibers of the graft molecular layer and star-shaped astrocytes of the graft granule cell layer and white matter. Normal amounts of vimentin immunoreactivity were seen in Bergmann fibers spanning the molecular layer and in astrocytes located in white matter areas of the transplants. Thus, the amount and distribution of GFAP and vimentin patterns of immunoreactivity suggested a rather normal astroglial development in the cerebellar grafts.[59,60]

In studies with E25 rabbit cerebellar grafts into the striatum or midbrain of newborn mice it was found by means of species-specific monoclonal anti-GFAP antibodies that astrocytes of donor origin migrate into the host CNS; moreover, the pattern of migration of transplant-derived astroglia or their precursors appears to be independent of the topographic origin of the transplant and therefore nonspecific toward defined regions of the host brain.[66]

Interestingly, when fetal rabbit striatal tissue is grafted into the posterior colliculus of newborn mice, astrocytes of donor origin migrate into the cerebellum of the host and at four weeks after transplantation present forms similar to the local glia, having transformed into, e.g., radial-like glia, which are not present in the striatum;[67] such observations support the idea that glial precursor cells are highly plastic, and that their form is defined by local conditions.[67]

During migration of implanted +/+ Purkinje cells into the cerebellar cortex of *pcd/pcd* mice (cf. corresponding section in chapter 7), a transient radial migration of the somata of host Golgi epithelial cells (the cells of origin of Bergmann glia) has been reported to take place from the interface between the lower part of the molecular layer and Purkinje cell layer, where they are normally located, to superficial sites of the molecular layer.[68] Such plastic changes may exert a facilitating role on the migratory process of donor Purkinje cells into the host cerebellum.

Tsurushima et al[38,39] found transient expression of tenascin, a neuron-glia cell adhesion molecule, in the grafted site 2-4 weeks after transplanting E14 Fischer 344 rat cerebellar tissue into the cerebellum of 8-week-old isogeneic hosts. Tenascin immunolabeling was detected transiently in radial glia processes adjacent to migratory Purkinje cells, mimicking a similar pattern of expression during normal cerebellar development and suggesting a possible involvement in the guidance of grafted neuron migration.

In an experiment designed to determine whether Bergmann fibers guiding the migration of transplanted Purkinje cells belong to donor or host tissue, Sotelo et al[69] implanted E12 cerebellar grafts from the Krox-20/lacZ14 transgenic mouse line into the cerebellum of C57BL/cdj *pcd/pcd* mice, so that Golgi epithelial cells and their Bergmann glial fibers in the donor (transgenic) tissue could be identified based on the expression of β-galactosidase activity. The results showed that 1-2 months after transplantation, β-galactosidase-positive glia are localized exclusively inside the graft, while Bergmann fibers of the host cerebellar cortex are β-galactosidase-negative, thus suggesting that glial guidance axes employed by donor Purkinje cells in their migratory process belong to the host. Further studies with E12 C57BL/6J +/+ grafts into C57BL/cdj *pcd/pcd* hosts, analyzed at 5, 7 and 13 days after grafting show that during the radial migration of donor Purkinje cells into the host cerebellar cortex, the involved Bergmann fibers of the host transiently re-express nestin (identified by means of Rat-401 immunoreactivity), a protein normally expressed by immature glia during stages of neuronal migration in the developing rat CNS.[69] It appears, therefore, that embryonic Purkinje cells may induce in the adult cerebellum a "rejuvenation" of host glia and the corresponding molecular plasticity needed for engraftment of the donor tissue.[69]

Seil[70] has proposed an astrocyte-mediated synapse-reduction mechanism for circuit reorganization after transplantation or in normal development, based on the finding that in the absence of functional glia there is a greater persistence of heterotypical synapses between recurrent axon collaterals and Purkinje dendritic spines in neonatal mouse cerebellar cultures.

GRAFT-HOST INTERACTIONS

Sympathetic adrenergic fibers from the host iris grow and functionally innervate intraocular cerebellar grafts in the rat.[71] Cerebellar grafts of E20-E25 fetal rabbit cerebellum, transplanted into the anterior eye chamber of athymic nude rats, receive excitatory cholinergic innervation from the host parasympathetic iris ground plexus, as well as sparse innervation of tyrosine hydroxylase positive fibers from the sympathetic plexus of the host iris.[50]

Serotonin-immunoreactive fibers grow from the adult host forebrain (fornix-fimbria, neocortex and hippocampus) into heterotopic cerebellar transplants placed into the parietal neocortex.[44] Heterotopic grafts of rat fetal cerebellar tissue into the cerebral ventricles or lateral hypothalamus have also been reported to receive peptidergic input from the host brain, based on oxytocin and neurophysin fiber immunostaining.[52]

Neural grafting studies in adult rats with kainic acid lesions of the cerebellum[72] have shown that donor Purkinje cells preferentially invade regions of the host molecular layer that are devoid of host Purkinje cells; there, they receive organotypic climbing fiber afferents that are organized along normal projectional maps, based on the anterograde transport of [³H]amino acids injected into the inferior olivary complex of the host.

Adult climbing fibers of normal rat cerebellum, labeled by means of *Phaseolus vulgaris* leucoagglutinin (PHA-L), can be induced by fetal cerebellar grafts to sprout new collaterals that terminate on donor Purkinje cells.[73] The phenomenon seems to be specifically elicited by fetal cerebellar grafts, as neocortical tissue grafts have no such effect. On the other hand, transplants of medullary embryonic tissue, containing the primordial inferior olivary complex, into the cerebellum of rats previously subjected to 3-acetylpyridine lesions of the endogenous olivocerebellar projection lead to synaptic formation between donor climbing fiber terminals and host Purkinje cells.[74]

Cerebellar tissue has been transplanted into the spinal cord of dogs to repair experimental transection injury.[75] In another study, cerebellar grafts inserted into the hemisected spinal cord of Sprague-Dawley rats at the T8 segment were found to rescue axotomized Clarke's nucleus neurons at the L1 level, which normally project to the cerebellum, from cellular death, although the somatic atrophy could not be prevented.[76]

Efferent projections from axons of heterotopically transplanted Purkinje cells, immunocytochemically labeled with anti-Leu-4 (CD3) antibody, into the rat striatum have been reported over distances of at least 500 µm at 2.5 months after grafting.[33] In a different experimental setting, axons of transplanted olfactory bulb neurons (marked by BALB/c strain of mouse allelic form of Thy-1.2 antibodies) invade the host cerebellum and elongate into the granule cell layer of the host (marked by the AKR strain of Thy-1.1), where they form asymmetrical synapses with local dendrites, most likely of host granule cells.[77] The granule cell layer is also able to receive novel "retinocerebellar" synapses from regenerating retinal ganglion cell axons guided to the cerebellum of adult hamsters by means of a peripheral nerve graft.[78]

REFERENCES

1. Cotman CW, ed. Synaptic Plasticity. New York: Guilford, 1985.
2. Björklund A. Intracerebral transplantation: prospects for neuronal replacement in neurodegenerative diseases. Res Publ Assoc Res Nerv Ment Dis 1993; 71:361-374.
3. Dunnett SB, Björklund A, eds. Functional Neural Transplantation. New York: Raven Press, 1994.
4. Allerand CD. Patterns of neuronal differentiation in developing cultures of neonatal mouse cerebellum: A living and silver impregnation study. J Comp Neurol 1971; 142:167-204.
5. Herndon RM, Seil FJ, Seidman C. Synaptogenesis in mouse cerebellum: A comparative in vivo and tissue culture study. Neuroscience 1981; 6:2587-2598.
6. Seil FJ. Cerebellum in tissue culture. Rev Neurosci 1979; 4:105-177.
7. Das GD, Altman J. Transplanted precursors of nerve cells: Their fate in the cerebellums of young rats. Science 1971; 173:637-638.
8. Das GD, Altman J. Studies on the transplantation of developing neural tissues in the mammalian brain. I. Transplantation of cerebellar slabs into the cerebellum of neonate rats. Brain Res 1972; 38:233-249.
9. Das GD. Transplantation of cerebellar tissue in the cerebellum of neonate rabbits. Brain Res 1973; 50:170-173.
10. Hine RJ. Transplanted cerebellar tissue in the rat: Its growth and its afferents. Anat Rec 1977; 187:605.
11. Wells J, McAllister JP. The development of cerebellar primordia transplanted to the neocortex of the rat. Dev Brain Res 1982; 4:167-179.
12. Alvarado-Mallart RM, Sotelo C. Differentiation of cerebellar anlage heterotopically transplanted to adult rat brain: A light and electron microscopic study. J Comp Neurol 1982; 212:247-267.
13. Kromer LF, Björklund A, Stenevi U. Intracephalic implants: A technique for studying neuronal interactions. Science 1979; 204: 1117-1119.
14. Kromer LF, Björklund A, Stenevi U. Intracephalic embryonic neural implants in the adult rat brain. I. Growth and mature organization of brainstem, cerebellar, and hippocampal implants. J Comp Neurol 1983; 218:433-459.
15. Ezerman EB, Kromer LF. Development and neuronal organization of dissociated and reaggregated embryonic cerebellum after intracephalic transplantation to adult rodent recipients. Dev Brain Res 1985; 23:287-292.
16. Ezerman EB. Survival and development of embryonic and postnatal cerebellum transplanted into adult rat hosts: effect of growth as explants in culture prior to transplantation. Dev Brain Res 1988; 41:253-261.

17. Alexandrova MA, Polezhaev LV. Transplantation of various regions of embryonic brain tissue into the brain of adult rats. J Hirnforsch 1984; 25:89-98.

18. Kikuchi Y. Transplantation of embryonic cerebella into adult rat cerebella. Brain Nerve (Tokyo) 1989; 41:45-53.

19. Chopko BW, Voneida TJ. Fetal rat cerebellar fragment transplantation into adult rat forebrain lesion cavities. J Neural Transpl Plast 1992; 3:63-69.

20. Stenevi U, Björklund A, Svendgaard N-A. Transplantation of central and peripheral monoamine neurons to the adult rat brain: Techniques and conditions for survival. Brain Res 1976; 114:1-20.

21. Takács J, Tran Minh Nhon T, Hámori J. Electron microscopical study of synaptic glomeruli in cerebellum transplanted to the anterior eye chamber. Acta Biol Hung (Budapest) 1986; 37:259-276.

22. Woodward DJ, Seiger Å, Olson L et al. Intrinsic and extrinsic determinants of dendritic development as revealed by Golgi studies of cerebellar and hippocampal transplants *in oculo.* Exp Neurol 1977; 57:984-998.

23. Takács J, Hámori J. Morphological study of cerebellar transplant cocultivated with cerebral cortical graft in the anterior eye chamber. I. Granular layer. Anat Embryol (Berl) 1988; 177:543-556.

24. Hámori J, Takács J. Morphological study of cerebellar transplant cocultivated with cerebral cortical graft in the anterior eye chamber. II. Purkinje cells and molecular layer. Anat Embryol (Berl) 1988; 177:557-569.

25. Wiestler OD, Aguzzi A, Schneemann M et al. Oncogene complementation in fetal brain transplants. Cancer Res 1992; 52:3760-3767.

26. Wiestler OD, Brustle O, Eibl RH et al. Retrovirus-mediated oncogene transfer into neural transplants. Brain Pathol 1992; 2:47-59.

27. Snyder EY, Deitcher DL, Walsh C et al. Multipotent neural cell lines can engraft and participate in development of mouse cerebellum. Cell 1992; 68:33-51.

28. Tsuda M, Yuasa S, Fujino Y et al. Retrovirus-mediated gene transfer into mouse cerebellar primary culture and its application to the neural transplantation. Brain Res Bulletin 1990; 24:787-792.

29. Yuasa S, Tsuda M, Kawamura K. Fate and behavior of genetically labeled cerebellar cells after transplantation into mouse cerebellum. Neurosci Res 1993; 17:257-263.

30. Anagnostopoulos J, Knoth R, Duffner T et al. Vascularization of fetal cerebellar tissue transplanted into the striatum of rats. In: Cervós-Navarro J, Ferszt R, eds. Stroke and Microcirculation. New York: Raven Press, 1987:171-176.

31. Kiessling M, Mies G, Paschen W et al. Blood flow and metabolism in heterotopic cerebellar grafts during hypoglycemia. Acta Neuropathol (Berl) 1988; 77:142-151.
32. Kleihues P, Kiessling M, Thilmann R et al. Resistance to hypoglycemia of cerebellar transplants in the rat forebrain. Acta Neuropathol (Berl) 1986; 72:23-28.
33. Gerloff C, Knappe UJK, Hettmannsperger U et al. Intrastriatal cerebellar grafts: differentiation of cerebellar anlage and sprouting of Purkinje cell axons. Dev Brain Res 1993; 74:30-40.
34. Rosenstein JM. Permeability to blood-borne protein and [^{3}H]GABA in CNS tissue grafts. I. Intraventricular grafts. J Comp Neurol 1991; 305:676-690.
35. Sotelo C. Cerebellar synaptogenesis: Mutant mice—neuronal grafting. J Physiol (Paris) 1991; 85:134-144.
36. Sotelo C. Cell interactions underlying Purkinje cell replacement by neural grafting in the *pcd* mutant cerebellum. Can J Neurol Sci 1993; 20 [Suppl 3]:S43-S52.
37. Sotelo C, Alvarado-Mallart RM, Gardette R et al. Fate of grafted embryonic Purkinje cells in the cerebellum of the adult "Purkinje cell degeneration" mutant mouse. I. Development of reciprocal graft-host interactions. J Comp Neurol 1990; 295:165-187.
38. Tsurushima H, Yuasa S, Kawamura K et al. Migration of donor Purkinje cells in the host adult rat cerebellum. Brain Nerve (Tokyo) 1993; 45:255-262.
39. Tsurushima H, Yuasa S, Kawamura K et al. Expression of tenascin and BDNF during the migration and differentiation of grafted Purkinje and granule cells in the adult rat cerebellum. Neurosci Res 1993; 18:109-120.
40. Marshak TL, Kleshchinov VN, Brodsky VYA. DNA content in neurons of embryonic cerebellum transplanted into the brain cortex of the adult rats. Ontogenez (Mosk) 1993; 24:53-57.
41. Celio MR, Baier W, Schärer L et al. Monoclonal antibodies directed against the calcium binding protein Calbindin D-28k. Cell Calcium 1990; 11:599-602.
42. Sotelo C, Alvarado-Mallart RM. Growth and differentiation of cerebellar suspensions transplanted into the adult cerebellum of mice with heredodegenerative ataxia. Proc Natl Acad Sci USA 1986; 83:1135-1139.
43. Triarhou LC, Low WC, Ghetti B. Intraparenchymal grafting of cerebellar cell suspensions to the deep cerebellar nuclei of *pcd* mutant mice, with particular emphasis on re-establishment of a Purkinje cell cortico-nuclear projection. Anat Embryol (Berl) 1992; 185:409-420.
44. Sotelo C, Alvarado-Mallart RM. Cerebellar transplants: Immuno-

cytochemical study of the specificity of Purkinje cell inputs and outputs. In: Björklund A, Stenevi U, eds. Neural Grafting in the Mammalian CNS. Amsterdam: Elsevier, 1985:205-215.

45. Chang AC, Triarhou LC, Alyea CJ et al. Developmental expression of polypeptide PEP-19 in cerebellar suspensions transplanted into the cerebellum of *pcd* mutant mice. Exp Brain Res 1989; 76:639-645.

46. Rouse RV, Sotelo C. Grafts of dissociated cerebellar cells containing Purkinje cell precursors organize into zebrin I defined compartments. Exp Brain Res 1990; 82:401-407.

47. Wassef M, Sotelo C, Thomasset M et al. Expression of compartmentation antigen zebrin I in cerebellar transplants. J Comp Neurol 1990; 294:223-234.

48. Sotelo C. Transplantation de neurones embryonnaires dans le cervelet de souris: Restauration de l' intégrité cérébelleuse chez des souris avec ataxie hérédo-dégénérative. Méd Sci 1988; 8:507-514.

49. Tokunaga A, Ono K, Date I et al. A monoclonal antibody that labels Purkinje cells in the rat cerebellum. Brain Res Bulletin 1991; 27:669-674.

50. Hall M, Wang Y, Granholm AC et al. Comparison of fetal rabbit brain xenografts to three different strains of athymic nude rats: electrophysiological and immunohistochemical studies of intraocular grafts. Cell Transpl 1992; 1:71-82.

51. Poltorak M, Freed WJ, Sternberger LA et al. A comparison of intraventricular and intraparenchymal cerebellar allografts in rat brain: evidence for normal phosphorylation of neurofilaments. J Neuroimmunol 1988; 20:63-72.

52. Perlow MJ, Nilaver G, Beinfeld MC et al. Host-graft interactions following cerebellar transplantation in rat. Soc Neurosci Abstr 1984; 10:663.

53. Wenthold RJ, Yokotani N, Doi K et al. Immunochemical characterization of the non-NMDA glutamate receptor using subunit-specific antibodies: Evidence for a hetero-oligomeric structure in rat brain. J Biol Chem 1992; 267:501-507.

54. Triarhou LC, Zhang W, Lee W-H. Graft-induced restoration of function in hereditary cerebellar ataxia. Neuroreport 1995; 6:1827-1832.

55. Triarhou LC, Zhang W, Lee W-H. Amelioration of the behavioral phenotype in genetically ataxic mice through bilateral intracerebellar grafting of fetal Purkinje cells. Cell Transpl 1996; 5:269-277.

56. Stasi K, Mitsacos A, Triarhou LC, Kouvelas ED. Functional integration of transplanted Purkinje cells into the atrophic cerebellum: I. Excitatory amino acid receptors and afferent innervation. Abstr Am Soc Neural Transpl 1996; 3:50.

57. Zhang W, Lee W-H, Triarhou LC. Grafted cerebellar cells in a mouse model of hereditary ataxia express IGF-I system genes and partially restore behavioral function. Nature Med 1996; 2:65-71.
58. Hoffer BJ, Seiger Å, Ljungberg T et al. Electrophysiological studies of brain homografts in the anterior chamber of the eye: Maturation of cerebellar cortex *in oculo*. Brain Res 1974; 79:165-184.
59. Björklund H, Bickford P, Dahl D et al. Intracranial cerebellar grafts: Intermediate filament immunohistochemistry and electrophysiology. Exp Brain Res 1984; 55:372-385.
60. Björklund H, Bickford P, Dahl D et al. Morphological and functional properties of intracranial cerebellar grafts. In: Björklund A, Stenevi U, eds. Neural Grafting in the Mammalian CNS. Amsterdam: Elsevier, 1985:191-203.
61. Kawamura K, Nanami T, Kikuchi Y et al. Grafted granule and Purkinje cells can migrate into the mature cerebellum of normal adult rats. Exp Brain Res 1988; 70:477-484.
62. Rossi F, Borsello T, Strata P. Embryonic Purkinje cells grafted on the surface of the cerebellar cortex integrate in the adult unlesioned cerebellum. Eur J Neurosci 1992; 4:589-594.
63. Rossi F, Borsello T. Ectopic Purkinje cells in the adult rat: Olivary innervation and different capabilities of migration and development after grafting. J Comp Neurol 1993; 337:70-82.
64. Mugnaini E, Morgan JI. The neuropeptide cerebellin is a marker for two similar neuronal circuits in rat brain. Proc Natl Acad Sci USA 1987; 84:8692-8696.
65. Alvarez-Otero R, Sotelo C, Alvarado-Mallart RM. Chick/quail chimeras with partial cerebellar grafts: An analysis of the origin and migration of cerebellar cells. J Comp Neurol 1993; 333:597-615.
66. Jacque C, Suard I, Collins P et al. Migration patterns of donor astrocytes after reciprocal striatum-cerebellum transplantation into newborn hosts. J Neurosci Res 1991; 29:421-428.
67. Jacque C, Tchelingerian JL, Collins P et al. In situ transformation of striatal glia into cerebellar-like glia after brain transplantation. Neurosci Lett 1992; 136:181-184.
68. Chang AC, Ghetti B. Embryonic cerebellar graft development during acute phase of gliosis in the cerebellum of *pcd* mutant mice. Chin J Physiol (Taipei) 1993; 36:141-149.
69. Sotelo C, Alvarado-Mallart RM, Frain M et al. Molecular plasticity of adult Bergmann fibers is associated with radial migration of grafted Purkinje cells. J Neurosci 1994; 14:124-133.
70. Seil FJ. Persistence of heterotypical synapses in transplanted cerebellar cultures in the absence of functional glia. Int J Dev Neurosci 1994; 12:411-421.
71. Hoffer BJ, Olson L, Seiger Å et al. Formation of a functional

adrenergic input to intraocular cerebellar grafts: Ingrowth of sympathetic fibers and inhibition of Purkinje cell activity by adrenergic input. J Neurobiol 1975; 6:565-586.

72. Armengol JA, Sotelo C, Angaut P et al. Organization of host afferents to cerebellar grafts implanted into kainate lesioned cerebellum of adult rats: Hodological evidence for the specificity of host-graft interactions. Eur J Neurosci 1989; 1:75-93.

73. Rossi F, Borsello T, Strata P. Embryonic Purkinje cells grafted on the surface of the adult uninjured rat cerebellum migrate in the host parenchyma and induce sprouting of intact climbing fibres. Eur J Neurosci 1994; 6:121-136.

74. Kawamura K, Murase S, Yuasa S et al. Transplantation of embryonic olive in the climbing-fiber-deprived adult rat cerebellum: Synaptogenesis on host Purkinje dendritic spines by donor climbing fibers. Neurosci Res [Suppl] 1990; 13:S61-S64.

75. Nan LZ. Experimental study on embryo cerebellar tissue transplantation in repairing transection injury of spinal cord. Chin J Surg (Beijing) 1989; 27:247-249.

76. Himes BT, Goldberger ME, Tessler A. Grafts of fetal central nervous system tissue rescue axotomized Clarke's nucleus neurons in adult and neonatal operates. J Comp Neurol 1994; 339:117-131.

77. Fujii M, Hayakashi T. Axons from the olfactory bulb transplanted into the cerebellum form synapses with dendrites in the granular layer, as demonstrated by mouse allelic form of Thy-1 and electron microscopy. Neurosci Res 1992; 14:73-78.

78. Zwimpfer TJ, Aguayo A, Bray GM. Synapse formation and preferential distribution in the granule cell layer by regenerating retinal ganglion cell axons guided to the cerebellum of adult hamsters. J Neurosci 1992; 12:1144-1159.

STRUCTURAL INTEGRATION OF CEREBELLAR GRAFTS IN ATAXIC MOUSE MUTANTS

INTRODUCTION

There are many mutations in the laboratory mouse that interfere with the formation and maintenance of the cerebellar circuitry (refs. 1, 2 and chapter 4 in this book). The cerebellar lesion may consist in either defective positioning of specific neuronal populations or selective loss. Such mutations provide unique material for investigating developmental and degenerative events because: *(i)* the background on the cellular architecture and synaptic connections of the cerebellum is strong; *(ii)* the cerebellum is a relatively simple neuronal circuit for studying phenomena with general implications for the CNS and *(iii)* the molecular genetics and chromosomal structure have been characterized in the laboratory mouse better than in any other mammal. In addition to insight into cerebellar ontogeny, neurological mutants offer invaluable experimental models pertinent to the neuropathological lesions of the human cerebellar ataxias. A description of cerebellar transplantation studies in cerebellar mutants (Table 7.1) addressing various issues follows.

TRANSPLANTATION STUDIES IN STAGGERER MUTANT MICE

The staggerer *(sg)* mutation is autosomal recessive. Purkinje cells are reduced in number and have abnormal dendritic branches that

Table 7.1. Diagrammatic summary of neurological mutant mice used as models for cerebellar transplantation experiments.

Mutant mouse	Gene symbol	Major deficit	Donor tissue	Host type	Refs.
Staggerer	sg	Inadequate formation of Purkinje-granule cell synapses; loss of Purkinje and granule cells	Mutant and normal fetal cerebellum	Normal mice	10
Weaver	wv	Loss of granule and Purkinje cells	Normal or mutant, fetal or early postnatal cerebellum	Mutant or normal mice	14-20
Purkinje cell degeneration	pcd	Loss of Purkinje cells	Normal fetal cerebellum	Mutant mice	15-17, 31-56
Nervous	nr	Loss of Purkinje cells	Normal fetal cerebellum	Mutant mice	44
Lurcher	Lc	Loss of Purkinje cells	Normal fetal cerebellum	Mutant mice	61,62

lack the peripheral components, i.e., the spiny branchlets, leading to inability of synapse formation between parallel fiber nerve endings and Purkinje cell spines, and eventually causing progressive degeneration of granule cells during the third and fourth weeks of life.[3,4] The carbohydrate pattern on staggerer cerebellar cell surface remains immature,[5,6] and the regulation of certain oligosaccharide hydrolyzing enzymes is abnormal.[7,8] Further, a disturbance in the conversion of the embryonic form of the neural cell adhesion molecule into the adult form has been described in staggerer mutants.[9]

Pieces of embryonic day 11 (E11) cerebellar anlagen from *sg*/+ × *sg*/+ matings were transplanted into the anterior eye chamber of wild-type recipient mice, aiming at studying the character of the *sg* mutation.[10] Six weeks after transplantation in oculo, graft viability was 80%. About 35% of the surviving transplants contained Purkinje, Golgi and deep nuclei macroneurons, but no or very few granule cells, a proportion within the range of the 1:2:1 genetic probability of *sg/sg* : *sg/+* : *+/+* donor tissue genotype, and consistent with an intrinsic action of the *sg* gene in determining the phenotype of the transplanted tissue. On the other hand, wild-type grafts maturing in the eye of wild-type hosts contained granule cells, as well as macroneurons, in 100% of the cases.[10]

CEREBELLAR TRANSPLANTATION IN WEAVER MUTANT MICE

The weaver *(wv)* mutation (mouse chr. 16) leads to massive death of postmitotic granule cell precursors during the first 15 days of postnatal life and to a reduced number of Purkinje cells in the cerebellum of homozygotes; heterozygotes also manifest similar phenotypic expressions albeit to a lesser extent, thus making the mutation incomplete dominant.[11-14]

Solid grafts of E15 wild-type cerebellar tissue were transplanted into the cerebellomedullary cistern of weaver hosts, between the uvula vermis and the dorsal surface of the brainstem, to study their survival, growth and synaptic properties inside the cerebrospinal fluid of the mutant environment.[14-17] The grafts displayed a layered cellular organization reminiscent of the normal cerebellar cortex, with identifiable molecular, Purkinje cell and granule cell layers. Parallel fiber axon terminals presynaptic to Purkinje cell dendritic spines were identified in the molecular layer of the grafts. The number of parallel fibers, however, was reduced compared to the normal cerebellar cortex, a phenomenon commonly seen in cerebellum, in tissue culture or in cerebellar transplants into normal hosts. It was concluded that the weaver environment does not pose any apparent limitations beyond those inherent in the process of cerebellar growth and differentiation outside its normal anatomical context.

In another study, pieces of E15 wild-type cerebellar tissue were transplanted into the cerebellum of 4-week-old weaver mutants.[18,19] Six weeks after transplantation, donor tissue developed a trilaminar organization, which contrasted with the granuloprival cerebellar cortex of the hosts. Evidence for the migration of implanted granule cells into the host cerebellum was presented. Positive immunreactivity for synapsin I, a synaptic vesicle membrane-specific phosphoprotein, was taken as an index of synapse formation by donor granule and Purkinje cells, possibly on host cerebellar neurons.

Weaver-into-normal cerebellar grafts have also been performed.[20] Mutant granule cell precursors were prepared from the external germinal layer of the cerebellum of postnatal day 5-6 (P5-P6) weaver animals and implanted into the cerebellum of P5 wild-type hosts. Three to six days after transplantation, some *wv/wv* transplanted cells displayed features of differentiated granule cells, e.g., parallel fiber extension, migration through the molecular and Purkinje cell layers, and extension of dendrites. The conclusion of that study was that the *wv* gene acts nonautonomously in vivo, and that local cell interactions required for granule cell migration may induce early steps in neuronal differentiation.[20] However, a number of surviving granule cells is observed in the weaver cerebellum anyway, and the holding view is that the *wv* gene acts intrinsically within the mutant neurons in causing their cellular death.[21]

CEREBELLAR TRANSPLANTS IN *PCD* MUTANT MICE

By far the most widely used model in cerebellar transplantation is the "Purkinje cell degeneration" neurological mutant of the laboratory mouse (gene symbol *pcd*, mouse chr. 13). The *pcd* mutation is autosomal recessive and is responsible for a virtually complete degeneration of Purkinje cells between P17 and P45, i.e., after the full maturation of the cerebellar circuitry.[11,22-25] In that respect, it is the only known mutant characterized by a virtually complete loss of Purkinje cells during adulthood. Thus, both the temporal pattern and the degree of Purkinje cell degeneration render the *pcd* model ideal for transplantation studies pertinent to the cerebellar ataxias of Purkinje cell type. Behaviorally, *pcd* homozygotes manifest an ataxic syndrome beginning at 3-4 weeks of age.

A moderate degree of nerve cell loss—in the order of ~20%—that is secondary (anterograde transsynaptic) to the genetically-determined loss of Purkinje cells is observed in the deep cerebellar nuclei of 10-month-old *pcd* mutants.[26] About 50% of neurons in the inferior olivary complex of *pcd* homozygotes degenerate as well in a retrograde transsynaptic manner in response to the loss of Purkinje cells.[27] An exponential decay of granule cells is observed between 17 and 600 days of age, which finally reaches a 90% loss.[28] On the other hand, monoaminergic (catecholaminergic and serotoninergic) afferents to the cerebellum persist even after the degeneration of their target Purkinje cells.[29,30] A detailed review of the primary and secondary structural and biochemical changes occurring in the *pcd* model can be found in ref. 25.

The *pcd* mutant mouse has been used during the past ten years as a model for neural grafting by two groups of investigators, that of C. Sotelo, R.M. Alvarado-Mallart and colleagues[31-47] and our own at Indiana University.[15-17,48-56] The foci of the questions addressed by the two groups differ slightly, although a certain degree of overlap in concepts is evident. The former group has studied issues of cerebellar cortical plasticity and reconstruction, whereas the nucleus of our own studies has been the confrontation between wild-type and mutant cells, the reconstruction of the corticonuclear projection and the recovery of function.

The first logical step in using a mutant mouse as host for the implantation of genetically normal (wild-type) cells is to ensure that the microenvironment of the mutated organism is permissive to donor tissue survival and growth. Several factors directly or secondarily related to the genetically-induced degenerative process in the mutant brain might theoretically interfere with the growth and differentiation of grafted tissue.[17] Therefore, the fate of E14-E15 cerebellar implants was studied after grafting into the cerebellomedullary cistern of adult *pcd/pcd* recipients.[15-17] The grafts exhibited a layered cellular organization reminiscent of normal cerebellum (Fig. 7.1), and surviving Purkinje cells displayed typical cytological features, indicating that the environment of the mutant hosts did not appear to pose any apparent limitations to the application of neural grafting techniques for the correction of the neurological deficit. In another study with embryonic cell suspension grafts into the cerebellum of *pcd* mutant mice, it was found that donor Purkinje cells survive in larger numbers when

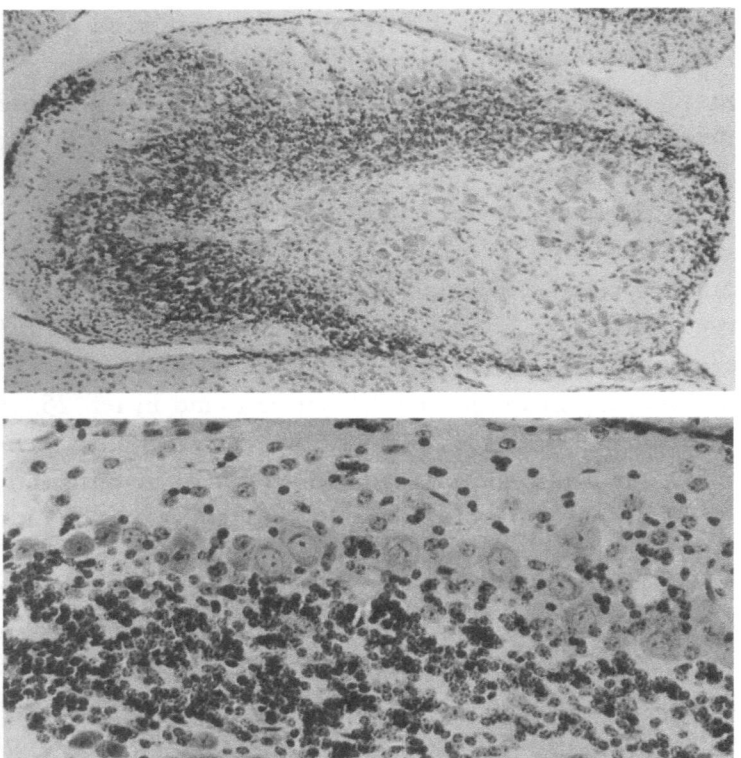

Fig. 7.1. Normal solid cerebellar grafts placed into the cerebellomedullary cistern of pcd mutant mice, 5.5 weeks after transplantation. Microscopic view of the entire graft (upper) lying between the host cerebellum and brainstem with the typical laminar organization. At higher power (lower), the cerebellar cortex of the grafted tissue contains both Purkinje and granule cells, as well as a well defined molecular layer. Ten micrometer thick paraffin sections stained with gallocyanin. Magnification x76 (upper), x300 (lower). Reprinted with permission from: Triarhou LC, Low WC, Ghetti B. Anat Embryol 1987; 176:145-154. © 1987 Springer-Verlag.

the transplantation is performed after the completion of the host degenerative process (i.e. after P45), as opposed to injecting the grafts during the ongoing degeneration of endogenous Purkinje cells, i.e., between P17 and P45.[49]

When E12 cerebellar grafts are implanted between two adjacent cerebellar cortical folia of *pcd* mutants in the form of either solid pieces or dissociated cell suspensions, donor Purkinje cells migrate along stereotyped pathways into the molecular layer of the deficient host cerebellum, where they develop flattened dendritic trees perpendicular to the host parallel fibers; donor Purkinje cell dendrites are composed of thick proximal branches and distal spiny branchlets that receive precisely segregated synaptic inputs from the adult host neuronal elements.[38-41] The timetable of these cellular interactions is remarkably similar to normal,[40] with one essential difference in the phase of radial migration, which occurs in an opposite direction, whereas during normal development the migration proceeds from the ventricular primitive neuroepithelium towards the cerebellar surface.[42] The developmental phases of Purkinje cell migration and dendritogenesis have been described thoroughly.[35,43,44,46] A positive neurotropism has been theorized to attract the grafted embryonic Purkinje cells into the host molecular layer.[31]

Transplanted Purkinje cells become synaptically integrated into the cerebellar circuitry of the deficient host brain by receiving afferent innervation from: *(i)* parallel fibers, as determined by electron microscopy;[38,41] *(ii)* climbing fibers, as determined by both electron microscopy and by electrophysiology of in vitro cerebellar slice preparations after juxtafastigial stimulation and intracellular recording from Purkinje cells[32,33,46] and *(iii)* serotoninergic axons, as determined by immunocytochemistry after selective neurotoxic removal of serotonin neurons from the cell suspensions prior to grafting.[53]

The main problem with cerebellar grafts placed into the cerebellar cortex of *pcd* hosts is the re-establishment of a corticonuclear Purkinje cell efferent projection. Such a difficulty has been attributed to a "physicochemical barrier" imposed by the host granule cell layer and white matter.[34,47] Aiming at reconstructing the corticonuclear projection, we transplanted cerebellar cell suspensions intraparenchymally into the deep cerebellar nuclei of *pcd* mutants.[51,52] Compared to "intracortical" grafts, the "intranuclear" grafting protocol features: *(i)* a new Purkinje axonal plexus innervating the host deep cerebellar nuclei, which is imperative in considering any form of functional improvement; *(ii)* a migratory process of Purkinje cell somata to the host cerebellar cortex that recapitulates the normal ontogenetic pattern and *(iii)* a correct orientation of dendritic trees toward the pia[50,52] (Figs. 7.2, 7.3 and 7.4). The functional effects of such intraparenchymal cerebellar grafts and some pathophysiological considerations are discussed in chapter 8.

Fig. 7.2. Intraparenchymal graft placement into the deep cerebellar nuclei of recipient pcd mutant mice. Immunocytochemical labeling with antibodies against 28 kDa Ca²⁺-binding protein (CaBP) reveals the distribution of donor Purkinje cells, since virtually all of the host Purkinje cells have been lost. At 5 days after grafting (upper), donor Purkinje cells are confined to a cluster (arrow) inside the host parenchyma; at 7 days after grafting (middle), a migratory stream of CaBP-immunoreactive cells is formed (double smaller arrows), starting at the cluster of the original graft site (single larger arrow), with a direction toward areas of the host cerebellar cortex; sagittal sections. At 30 days after grafting (lower), donor Purkinje cells cover a substantial part of the host, following the geometrical shape of the cerebellar cortex; coronal section with bilateral grafts. Magnification x40 (upper, middle), x12 (lower). Reprinted with permission from: Triarhou LC, Low WC, Ghetti B. Anat Embryol 1992; 185:409-420. © 1992 Springer-Verlag.

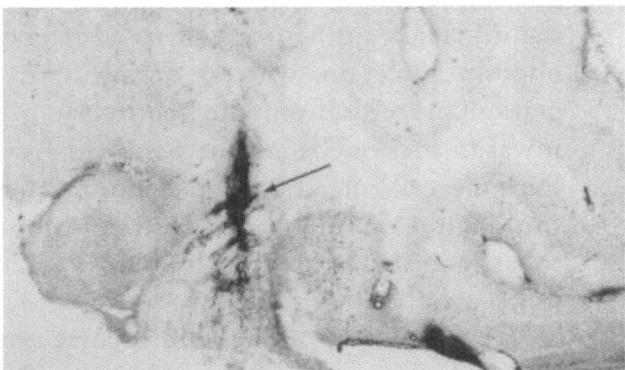

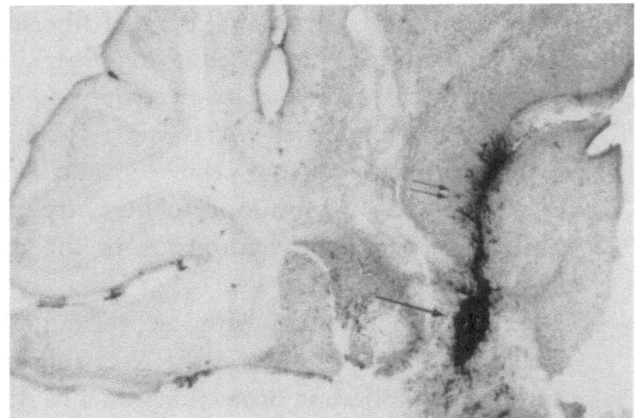

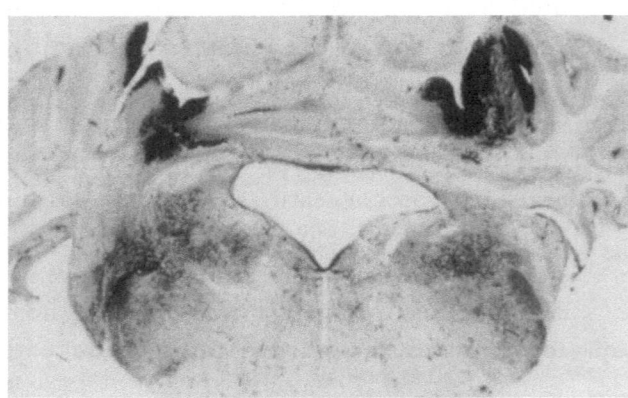

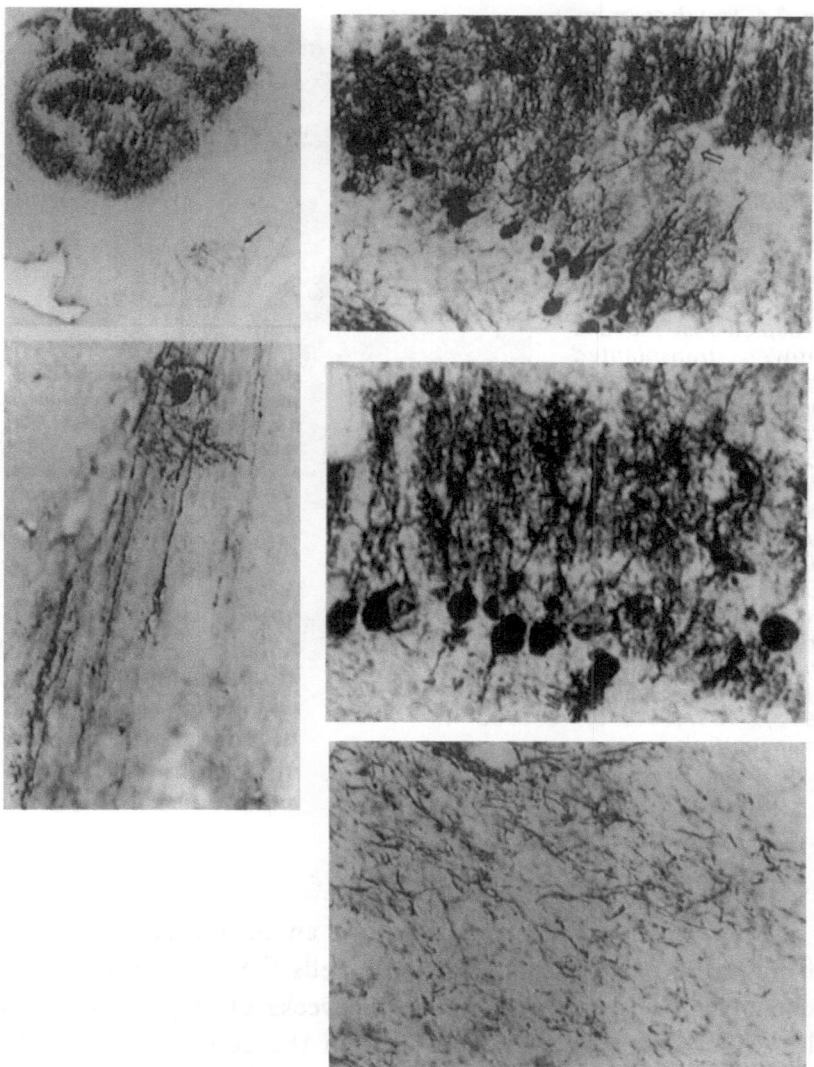

Fig. 7.3. Immunocytochemical correlates of cerebellar graft survival and outgrowth at 1.5 months after transplantation into the cerebella of pcd mutants. Alignment of grafted, 28 kDa Ca²⁺-binding protein (CaBP)-immunoreactive Purkinje cells along cortical folia and innervation of the deep cerebellar nuclei of the host by CaBP-immunopositive axons (upper left). Monoplanar disposition and correct orientation of the dendritic trees toward the pia in transplanted Purkinje cells settled in the host cerebellar cortex (right upper and middle). Axonal outgrowth of transplanted Purkinje cells, immunoreactive for CaBP (lower left). Innervation of the host deep cerebellar nuclear complex by a CaBP-immunoreactive nerve terminal plexus provided by grafted Purkinje cells. Magnification ×27 (upper left), ×160 (upper right, lower left), ×260 (middle and lower right). Micrograph in right middle unpublished. All others reprinted with permission from: Triarhou LC, Zhang W, Lee W-H. Neuroreport 1995; 6:1827-1832. © 1995 Rapid Science Publishers.

Fig. 7.4. Pseudocolor imaging of a parasagittal section from pcd cerebellum with an intraparenchymal cerebellar graft, immunocytochemically labeled for 28 kDa Ca²⁺-binding protein (CaBP), a selective marker for Purkinje cells, at one month's survival time. CaBP immunoreactivity is depicted in red and shows transplanted Purkinje cells, while immunonegativity, shown in blue, marks the host cerebellum. The graft occupies a little less than one-half of the host cerebellum at this particular level. Magnification x36. Unpublished computer-generated image, based on original material from: Triarhou LC, Low WC, Ghetti B. Anat Embryol 1992; 185:409-420. © 1992 Springer-Verlag.

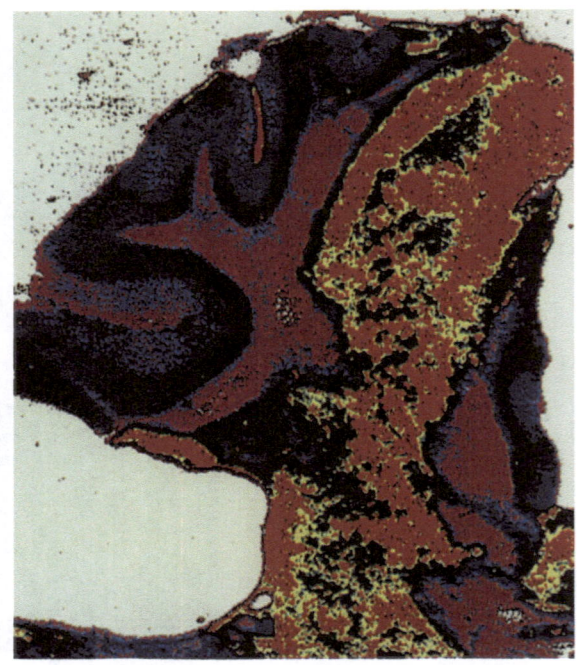

CEREBELLAR TRANSPLANTS IN NERVOUS MUTANT MICE

The nervous *(nr)* mutation follows an autosomal recessive trait of inheritance and affects Purkinje cells.[57,58] Most Purkinje cells degenerate between three and seven weeks of age. In two month old animals, 90% of Purkinje cells in the cerebellar hemispheres and 50% of Purkinje cells in the vermis have died off. Thus, compared with the *pcd* model, the degeneration of Purkinje cells in the nervous mutant is more protracted and less extensive.

In transplantation experiments, E12 cerebellar grafts were implanted into the intermediate cerebellar cortex of *nr/nr* mutants and allowed survival times of 2-4 months after grafting.[44] The general conclusion was that grafted Purkinje cells invade the host molecular layer with preference for regions of the host cerebellar cortex that are devoid of endogenous Purkinje cells, and stop their migration once they encounter the dendritic trees of host Purkinje cells.

CEREBELLAR TRANSPLANTS
IN LURCHER MUTANT MICE

The Lurcher *(Lc)* mutation (mouse chr. 6) is autosomal dominant and leads to extensive Purkinje cell death commencing around P8, i.e., during development; by one month of age Purkinje cell loss amounts to more than 90%, and by two months the Lurcher cerebellum is virtually devoid (i.e. by about 99%) of Purkinje cells.[11,59,60]

The Lurcher cerebellum has been used as a host model for the transplantation of cerebellar grafts prepared from wild-type donor mouse embryos in two independent studies.[61,62]

One study made use of E12 cerebellar cell suspensions implanted into the cerebellum of both juvenile (17 days old) and adult (1-6 months old) Lurcher mutants and survival times of 1-2 months after grafting.[61] The rate of graft survival in that study was 50% for both age groups of recipient mice. Purkinje cells from the grafts, immunolabeled with anti-CaBP antiserum, were found to infiltrate the atrophic cerebellar cortex of the host, occupying most frequently the molecular layer. The dendrites of the transplanted Purkinje cells failed in adopting the characteristic planar disposition inside the host cerebellum, an observation that was attributed to the severe depletion in the Lurcher mutant of granule cells and hence their parallel fibers,[60] elements that have a decisive role in the morphogenesis of the Purkinje cell dendritic tree during normal development.[63] Grafted Purkinje cells supplied an axonal innervation to the deep cerebellar nuclei of the hosts in 30% of the cases.[61]

Another study made use of E12-E14 solid cerebellar grafts implanted into the cerebellum of two- to six-month-old Lurcher recipients, with survival times of one to three months after grafting.[62] Donor Purkinje cells, immunoreactive for CaBP or cGMP-dependent protein kinase, were found to migrate into the granule cell and molecular layers of the host cerebellar cortex and to occasionally innervate the deep nuclear complex, but never in a massive fashion. A similar invasion by grafted Purkinje cells is also seen into the molecular layer of the host dorsal cochlear nucleus,[64] a brainstem structure that is anatomically homologous to the cerebellar cortex.[65] The dendritic trees of grafted Purkinje cells extend in the sagittal plane to some degree, but are not completely flat, again owing most likely to the marked parallel fiber deficit in the Lurcher cerebellum. An important finding of that study relates to the synaptic investment of grafted Purkinje cells, which is abnormal both in quantitative and qualitative terms. Synaptic inputs to both the perikaryon and dendrites of donor Purkinje cells are reduced; the compartmentation in proximal and distal dendritic segments is severely affected; climbing fiber afferents form synapses scarcely and, finally, large perisomatic baskets as well as "pinceau" formations around the axon initial segment are absent. Grafted Purkinje cells located in the host granule cell layer receive heterologous synapses from mossy fibers, a phenomenon previously observed in granuloprival cerebella.[13] In all, it seems that the restoration of the developmentally perturbed cerebellar circuit of the Lurcher mutant by means of neural transplantation poses certain limitations.

REFERENCES

1. Sidman RL. Mutations affecting the central nervous system in the mouse. In: Schmitt FO, Bird SJ, Bloom FE. Molecular Genetic Neuroscience. New York: Raven Press, 1982:389-400.
2. Sidman RL, Green MC, Appel SH. Catalog of the Neurological Mutants of the Mouse. Cambridge, MA: Harvard University Press, 1965.
3. Sidman RL, Lane PW, Dickie MM. Staggerer, a new mutation in the mouse affecting the cerebellum. Science 1962; 137:610-612.
4. Sotelo C, Changeux J-P. Transsynaptic degeneration 'en cascade' in the cerebellar cortex of staggerer mutant mice. Brain Res 1974; 67:519-526.
5. Hatten ME, Messer A. Postnatal cerebellar cells from staggerer mutant mice express embryonic cell surface characteristics. Nature (Lond) 1978; 276:504-506.
6. Trenkner E. Postnatal cerebellar cells of staggerer mutant mice express immature components on their surface. Nature (Lond) 1979; 277: 566-567.
7. Wille W, Heinlein UAO, Spier-Michl I et al. Development-dependent regulation of N-acetyl-β-D-hexosaminidase of cerebellum and cerebrum of normal and staggerer mutant mice. J Neurochem 1983; 40:235-239.
8. Wille W, Trenkner E. Changes in particulate neuraminidase activity during normal and staggerer mutant mouse development. J Neurochem 1981; 37:443-446.

9. Edelman GM, Chuong C-M. Embryonic to adult conversion of neural cell adhesion molecules in normal and staggerer mice. Proc Natl Acad Sci USA 1982; 79:7036-7040.

10. Wille W, Goldowitz D, Seiger Å et al. The neurological mutation staggerer is expressed in embryonic cerebellar transplants matured in the anterior eye chamber of normal mice. Neurosci Lett 1983; 42:1-6.

11. Davisson MT, Roderick TH. Linkage map. In: Lyon MF, Searle AG, eds. Genetic Variants and Strains of the Laboratory Mouse. 2nd ed. Oxford, Stuttgart: Oxford University Press, Gustav Fischer Verlag, 1989:416-427.

12. Rakic P, Sidman RL. Sequence of developmental abnormalities leading to granule cell deficit in cerebellar cortex of weaver mutant mice. J Comp Neurol 1973; 152:103-132.

13. Sotelo C. Mutant mice and the formation of cerebellar circuitry. Trends Neurosci 1980; 3:33-36.

14. Triarhou LC. Weaver gene expression in central nervous system. In: Conn PM, ed. Gene Expression in Neural Tissues. San Diego: Academic Press, 1992:209-227.

15. Triarhou LC, Ghetti B, Low WC. Purkinje and granule cells survive in cerebellar grafts implanted into hosts with genetically-determined Purkinje or granule cell degeneration. Ann Neurol 1986; 20:138.

16. Low WC, Triarhou LC, Ghetti B. Cerebellar transplants into mutant mice with Purkinje and granule cell degeneration. Ann NY Acad Sci 1987; 495:740-744.

17. Triarhou LC, Low WC, Ghetti B. Transplantation of cerebellar anlagen to hosts with genetic cerebellocortical atrophy. Anat Embryol (Berl) 1987; 176:145-154.

18. Takayama H, Kohsaka S, Shinozaki T et al. Immunohistochemical studies on synapse formation by embryonic cerebellar tissue transplanted into the cerebellum of the weaver mutant mouse. Neurosci Lett 1987; 79:246-250.

19. Takayama H, Toya S, Shinozaki T et al. Possible synapse formation by embryonic cerebellar tissue grafted into the cerebellum of the weaver mutant mouse. Acta Neurochir [Suppl] 1988; 43:154-158.

20. Gao W-Q, Hatten ME. Neuronal differentiation rescued by implantation of *weaver* granule cell precursors into wild-type cerebellar cortex. Science 1993; 260:367-370.

21. Goldowitz D. The weaver granuloprival phenotype is due to intrinsic action of the mutant locus in granule cells: Evidence from homozygous weaver chimeras. Neuron 1989; 2:1565-1575.

22. Mullen RJ, Eicher EM, Sidman RL. Purkinje cell degeneration, a new neurological mutation in the mouse. Proc Natl Acad Sci USA 1976; 73:208-212.

23. Mullen RJ. Site of *pcd* gene action and Purkinje cell mosaicism in the cerebella of chimeric mice. Nature (Lond) 1977; 270:245-247.

24. Landis SC, Mullen RJ. The development and degeneration of Purkinje cells in *pcd* mutant mice. J Comp Neurol 1978; 177:125-144.

25. Ghetti B, Triarhou LC. The Purkinje cell degeneration mutant: A model to study the consequences of neuronal degeneration. In: Plaitakis A, ed. Cerebellar Degenerations: Clinical Neurobiology. Boston: Kluwer Academic, 1992:159-181.

26. Triarhou LC, Norton J, Ghetti B. Anterograde transsynaptic degeneration in the deep cerebellar nuclei of Purkinje cell degeneration *(pcd)* mutant mice. Exp Brain Res 1987; 66:577-588.

27. Triarhou LC, Ghetti B. Stabilisation of neurone number in the inferior olivary complex of aged 'Purkinje cell degeneration' mutant mice. Acta Neuropathol (Berl) 1991; 81:597-602.

28. Triarhou LC, Norton J, Alyea C et al. A quantitative study of the granule cells in the Purkinje cell degeneration *(pcd)* mutant. Ann Neurol 1985; 18:146.

29. Triarhou LC, Ghetti B. Monoaminergic nerve terminals in the cerebellar cortex of Purkinje cell degeneration mutant mice: Fine structural integrity and modification of cellular environs following loss of Purkinje and granule cells. Neuroscience 1986; 18:795-807.

30. Triarhou LC, Ghetti B. Serotonin immunoreactivity in the cerebellum of two neurological mutant mice and the corresponding wild-type genetic stocks. J Chem Neuroanat 1991; 4:421-428.

31. Alvarado-Mallart RM, Sotelo C. Cerebellar grafting in murine heredodegenerative ataxia. Current limitations for a therapeutic approach. Adv Neurol 1993; 61:181-192.

32. Gardette R, Alvarado-Mallart RM, Crepel F, Sotelo C. Electrophysiological demonstration of a synaptic integration of transplanted Purkinje cells into the cerebellum of the adult Purkinje cell degeneration mutant mouse. Neuroscience 1988; 24:777-789.

33. Gardette R, Crepel F, Alvarado-Mallart RM et al. Fate of grafted embryonic Purkinje cells in the cerebellum of the adult "Purkinje cell degeneration" mutant mouse. II. Development of synaptic responses: An in vitro study. J Comp Neurol 1990; 295:188-196.

34. Keep M, Alvarado-Mallart RM, Sotelo C. New insights on the factors orienting the axonal outgrowth of grafted Purkinje cells in the *pcd* cerebellum. Dev Neurosci 1992; 14:153-165.

35. Sotelo C. Transplantation de neurones embryonnaires dans le cervelet de souris: Restauration de l' intégrité cérébelleuse chez des souris avec ataxie hérédo-dégénérative. Méd Sci 1988; 8:507-514.

36. Sotelo C. Cerebellar synaptogenesis: Mutant mice—neuronal grafting. J Physiol (Paris) 1991; 85:134-144.

37. Sotelo C. Cell interactions underlying Purkinje cell replacement by neural grafting in the *pcd* mutant cerebellum. Can J Neurol Sci 1993; 20 [Suppl 3]:S43-S52.

38. Sotelo C, Alvarado-Mallart RM. Growth and differentiation of cerebellar suspensions transplanted into the adult cerebellum of mice with heredodegenerative ataxia. Proc Natl Acad Sci USA 1986; 83:1135-1139.

39. Sotelo C, Alvarado-Mallart RM. Cerebellar transplantations in adult mice with heredo-degenerative ataxia. Ann NY Acad Sci 1987; 495:242-266.

40. Sotelo C, Alvarado-Mallart RM. Embryonic and adult neurons interact to allow Purkinje cell replacement in mutant cerebellum. Nature (Lond) 1987; 327:421-423.

41. Sotelo C, Alvarado-Mallart RM. Reconstruction of the defective cerebellar circuitry in adult Purkinje cell degeneration mutant mice by Purkinje cell replacement through transplantation of solid embryonic implants. Neuroscience 1987; 20:1-22.

42. Sotelo C, Alvarado-Mallart RM. Integration of grafted Purkinje cells into the host cerebellar ciruitry in Purkinje cell degeneration mutant mouse. Prog Brain Res 1988; 78:141-154.

43. Sotelo C, Alvarado-Mallart RM. The reconstruction of cerebellar circuits. Trends Neurosci 1991; 14:350-355.

44. Sotelo C, Alvarado-Mallart RM. Cerebellar grafting as a tool to analyze new aspects of cerebellar development and plasticity. In: Llinás R, Sotelo C, eds. The Cerebellum Revisited. New York-Berlin-Heidelberg: Springer-Verlag, 1992:84-115.

45. Sotelo C, Alvarado-Mallart RM, Frain M et al. Molecular plasticity of adult Bergmann fibers is associated with radial migration of grafted Purkinje cells. J Neurosci 1994; 14:124-133.

46. Sotelo C, Alvarado-Mallart RM, Gardette R et al. Fate of grafted embryonic Purkinje cells in the cerebellum of the adult "Purkinje cell degeneration" mutant mouse. I. Development of reciprocal graft-host interactions. J Comp Neurol 1990; 295:165-187.

47. Sotelo C, Alvarado-Mallart RM, Keep M. Fate of axons of embryonic Purkinje cells grafted in the adult cerebellum of the *pcd* mutant mouse. In: Letourneau PC, Kater SB, Macagno ER, eds. The Nerve Growth Cone. New York: Raven Press, 1992:505-517.

48. Chang AC, Triarhou LC, Alyea CJ et al. Developmental expression of polypeptide PEP-19 in cerebellar suspensions transplanted into the cerebellum of *pcd* mutant mice. Exp Brain Res 1989; 76:639-645.

49. Ghetti B, Triarhou LC, Alyea CJ et al. Timing of neuronal replacement in cerebellar degenerative ataxia of Purkinje cell type. Prog Brain Res 1990; 82:197-202.

50. Triarhou LC. Cerebellar transplantation in hereditary ataxia and the recovery of function: Why do the deep cerebellar nuclei represent a better graft site than the cerebellar cortex. Abstr Am Soc Neural Transpl 1995; 2:21.

51. Triarhou LC, Low WC, Ghetti B. Intraparenchymal grafting of cerebellar cell suspensions to the deep cerebellar nuclei of *pcd* mutant mice: Rationale and histochemical organization. Soc Neurosci Abstr 1989; 15:10.

52. Triarhou LC, Low WC, Ghetti B. Intraparenchymal grafting of cerebellar cell suspensions to the deep cerebellar nuclei of *pcd* mutant mice, with particular emphasis on re-establishment of a Purkinje cell cortico-nuclear projection. Anat Embryol (Berl) 1992; 185:409-420.

53. Triarhou LC, Low WC, Ghetti B. Serotonin fiber innervation of cerebellar cell suspensions intraparenchymally grafted to the cerebellum of *pcd* mutant mice. Neurochem Res 1992; 17:475-482.

54. Triarhou LC, Zhang W, Lee W-H. Graft-induced restoration of function in hereditary cerebellar ataxia. Neuroreport 1995; 6:1827-1832.

55. Triarhou LC, Zhang W, Lee W-H. Amelioration of the behavioral phenotype in genetically ataxic mice through bilateral intracerebellar grafting of fetal Purkinje cells. Cell Transpl 1996; 5:269-277.

56. Zhang W, Lee W-H, Triarhou, L.C. Grafted cerebellar cells in a mouse model of hereditary ataxia express IGF-I system genes and partially restore behavioral function. Nature Med 1996; 2:65-71.

57. Sidman RL, Green MC. 'Nervous', a new mutant mouse with cerebellar disease. In: Sabourdy M, ed. Les Mutants Pathologiques chez l' Animal. Paris: Éditions du Centre National de la Recherche Scientifique, 1970:69-79.

58. Landis SC. Ultrastructural changes in the mitochondria of cerebellar Purkinje cells of nervous mutant mice. J Cell Biol 1973; 57:782-797.

59. Phillips RJS. "Lurcher", a new gene in linkage group XI of the house mouse. J Genet 1960; 57:35-42.

60. Caddy KWT, Biscoe TJ. Structural and quantitative studies in the normal C3H and Lurcher mutant mouse. Phil Trans Roy Soc Lond (Biol) 1979; 287:167-201.

61. Tomey DA, Heckroth JA. Transplantation of normal embryonic cerebellar cell suspensions into the cerebellum of *Lurcher* mutant mice. Exp Neurol 1993; 122:165-170.

62. Dumesnil-Bousez N, Sotelo C. Partial reconstruction of the adult Lurcher cerebellar circuitry by neural grafting. Neuroscience 1993; 55:1-21.

63. Altman J. Morphological development of the rat cerebellum and

some of its mechanisms. Exp Brain Res [Suppl] 1982; 6:8-49.

64. Dumesnil-Bousez N, Sotelo C. The dorsal cochlear nucleus of the adult Lurcher mouse is specifically invaded by embryonic grafted Purkinje cells. Brain Res 1993; 622:343-347.

65. Mugnaini E, Morgan JI. The neuropeptide cerebellin is a marker for two similar neuronal circuits in rat brain. Proc Natl Acad Sci USA 1987; 84:8692-8696.

CEREBELLAR GRAFTING AND THE RECOVERY OF FUNCTION

INTRODUCTION

Evidence for functional recovery after cerebellar transplantation has been obtained in the *pcd* mouse model of hereditary cerebellar ataxia.[1-4] Grafts of embryonic day 11-12 (E11-E12) cerebellar cell suspensions were placed bilaterally into the deep cerebellar nuclei of the host mutants, according to the protocol that places emphasis on reconstructing the corticonuclear γ-aminobutyric acid (GABA)ergic projection.[5] Vehicle-injected *pcd* homozygotes were used as controls in the behavioral studies. Animals were tested in a battery of motor tasks six weeks postoperatively to determine the recovery of behavioral responses. Surviving Purkinje cells immunoreactive for 28 kDa Ca^{2+}-binding protein (CaBP) were found in all graft-recipient animals. Counts of CaBP-immunoreactive neurons in histochemical preparations of the transplanted cerebella, combined over both sides, yielded numbers in the range of 1000-6500 surviving Purkinje cells per animal, with a 2865 cell average.[2]

SPONTANEOUS MOVEMENT AND STANCE

Qualitative observations have disclosed that grafted *pcd* mice are able to keep their body in an upright posture, markedly contrasting with the lowered, widened stance of sham-operated mutants (Fig. 8.1); furthermore, they are capable of sustaining their

Fig. 8.1. Effects of bilateral cerebellar grafts into the deep cerebellar nuclei of pcd mutant mice on motor performance. Sham-injected mutant (upper), with the typical lowered, widened stance. Graft-recipient animal (lower), able to sustain its abdomen in a raised relief from the ground during movement. Reprinted with permission from: Triarhou LC, Zhang W, Lee W-H. Cell Transpl 1996; 5:269-277. © 1996 Elsevier Science Inc.

abdomen in a raised relief from the matrix floor and of moving about for relatively long periods of time without falling over; hindlimbs are less abducted and less hyperextended in transplant-receiving animals than in mice with vehicle injections.[2]

EQUILIBRIUM

As an index of equilibrium,[6-8] the time was measured between placement on and falling off a still balance rod suspended 13 cm over the ground.[3] Wild-type mice generally remain on the rod for long periods of time. Sham-injected *pcd* mutants stayed on the rod for an average of 2.7 sec; on the other hand, *pcd* mutant mice with bilateral cerebellar grafts stayed for an average 9.9 sec before falling off the bar, thus indicating a 3.6-fold improvement after transplantation.

MOTOR COORDINATION AND FATIGUE RESISTANCE

A rota-rod apparatus was used, designed for mice and rotating at 3 rpm (Fig. 8.2). The rota-rod treadmill paradigm is widely used to assess motor coordination and fatigue resistance in various brain abiotrophies.[9,10] Animals were tested one week preoperatively and six weeks postoperatively. Three successive trials were given to each mouse. Bilateral injections of vehicle did not appreciably modify the performance of *pcd* mutant mice in the rota-rod tests, while bilateral cerebellar grafts led to a 3.5-fold increase in the time period that *pcd* mutants stayed on the rotating drum based on the comparison of the three-trial mean scores, and to a 5.5-fold increase based on the comparison of the maximum scores out of the three trials[1] (Fig. 8.3). In particular, postoperative times on the rota-rod were as follows: for the sham-injected group, the average of the three-trial means was 3.0 sec, while the average of maximum scores was 4.2 sec; for the graft-receiving group, the average of the three-trial means was 13.5 sec, and the average of the maximum scores out of the three trials was 21.5 sec.[2]

OPEN-FIELD ACTIVITY

Quantification of motor activity was effected in a 5 × 5 square open-field matrix.[6] The pattern of animal movement was traced over an observation period of 5 minutes and the number of square-crossing events registered (Fig. 8.4). The tracking of movement paths showed wild-type mice to exhibit the most complex pattern of activity, sham-operated *pcd* mice the lowest activity and graft-recipient *pcd* mice being in between.[3] Normal animals display levels of activity in the vicinity of 200-250 square crossing events; overall activity is reduced to an average of 21.5 square crossings in sham-operated mutants, and an increase to an average 68.7 events is brought about after bilateral cerebellar transplants, which represents a 3.2-fold improvement in motor performance (Fig. 8.5).[2,3]

Fig. 8.2. Effects of bilateral cerebellar grafts into the deep cerebellar nuclei of pcd mutant mice on motor performance. Sham-injected mutant (upper) in the rota-rod treadmill apparatus as it is falling off the rotating drum only 5 sec after its placement on the rod. A graft-recipient animal (lower) that was able to stay on the apparatus for over 40 sec after placement on the rod. Reprinted with permission from: Triarhou LC, Zhang W, Lee W-H. Cell Transpl 1996; 5:269-277. © 1996 Elsevier Science Inc.

Fig. 8.3 (opposite). Rota-rod test results (means ± S.E.M.) in pcd mutant mice with sham injections (left groups, open circles) and bilateral cerebellar tissue grafts (right groups, open diamonds). Solid lines connect the mean scores in each group tested pre- and postoperatively. (a) Group means obtained from each animal's three-trial average score. (b) Group means obtained from each animal's maximum score from three trials. The graft-receiving pcd group differed significantly from the sham-operated pcd group in the contrast of mean scores (P=0.027) and maximum scores (P=0.016). Reprinted with permission from: Triarhou LC, Zhang W, Lee W-H. Neuroreport 1995; 6:1827-1832. © 1995 Rapid Science Publishers.

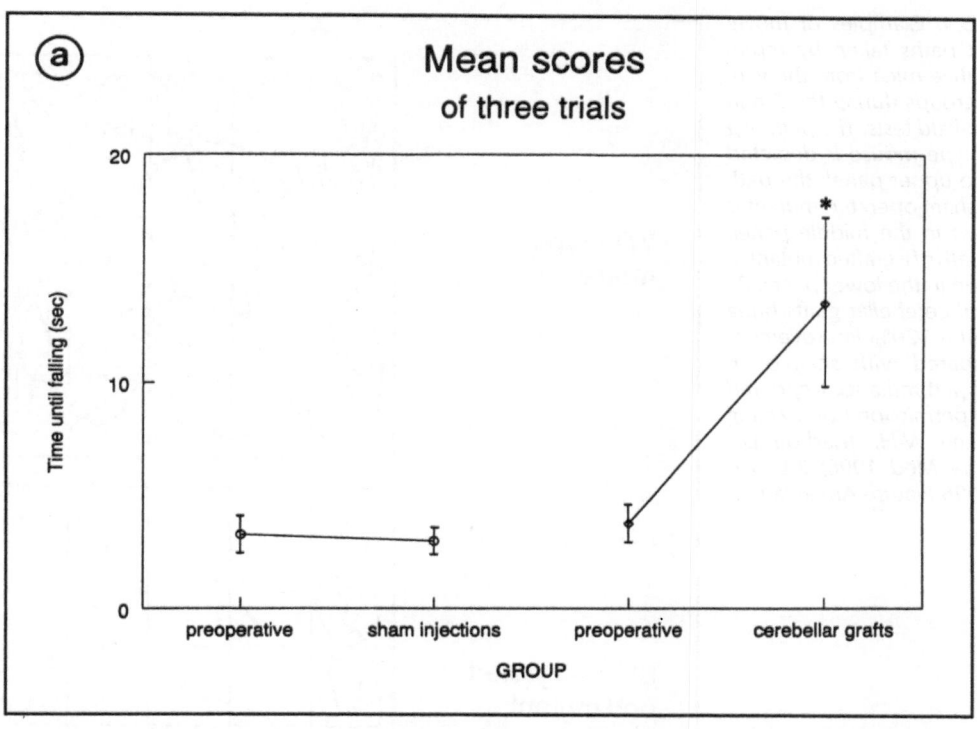

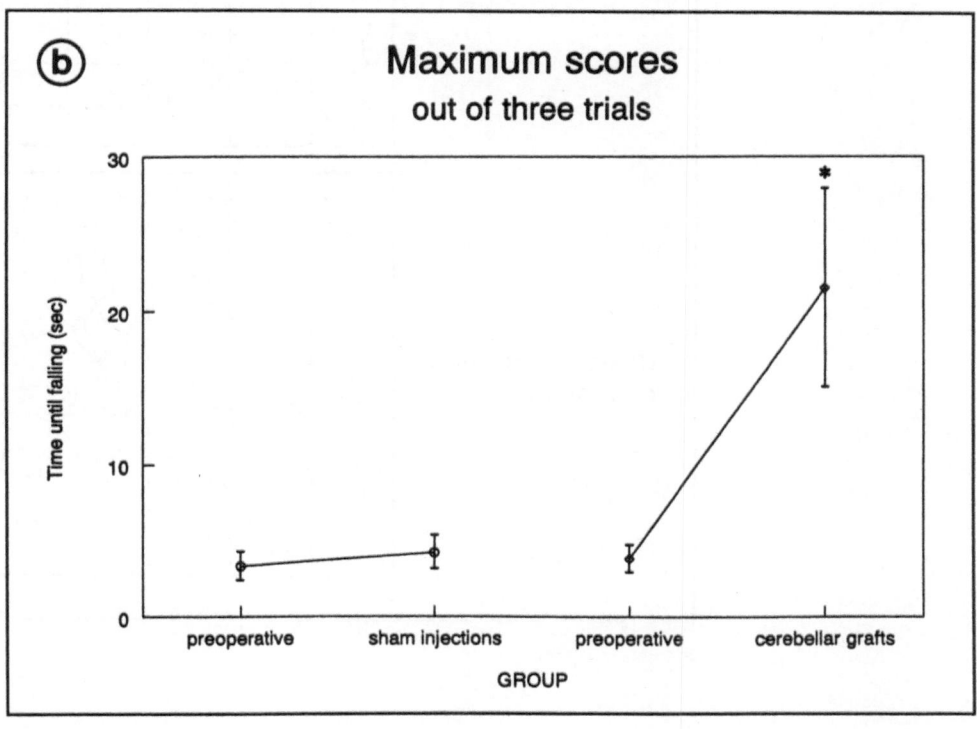

Fig. 8.4. Examples of move-ment paths taken by repre-sentative mice from the vari-ous groups during the 5 min open-field tests. The path of a wild-type mouse is depicted in the upper panel; the path of a sham-operated mutant is shown in the middle panel; the path of a grafted mutant in shown in the lower panel. Bi-lateral cerebellar grafts bring about a 220% improvement compared with sham-oper-ated pcd mutants. Reprinted with permission from: Zhang W, Lee W-H, Triarhou LC. Nature Med 1996; 2:65-71. © 1996 Nature America Inc.

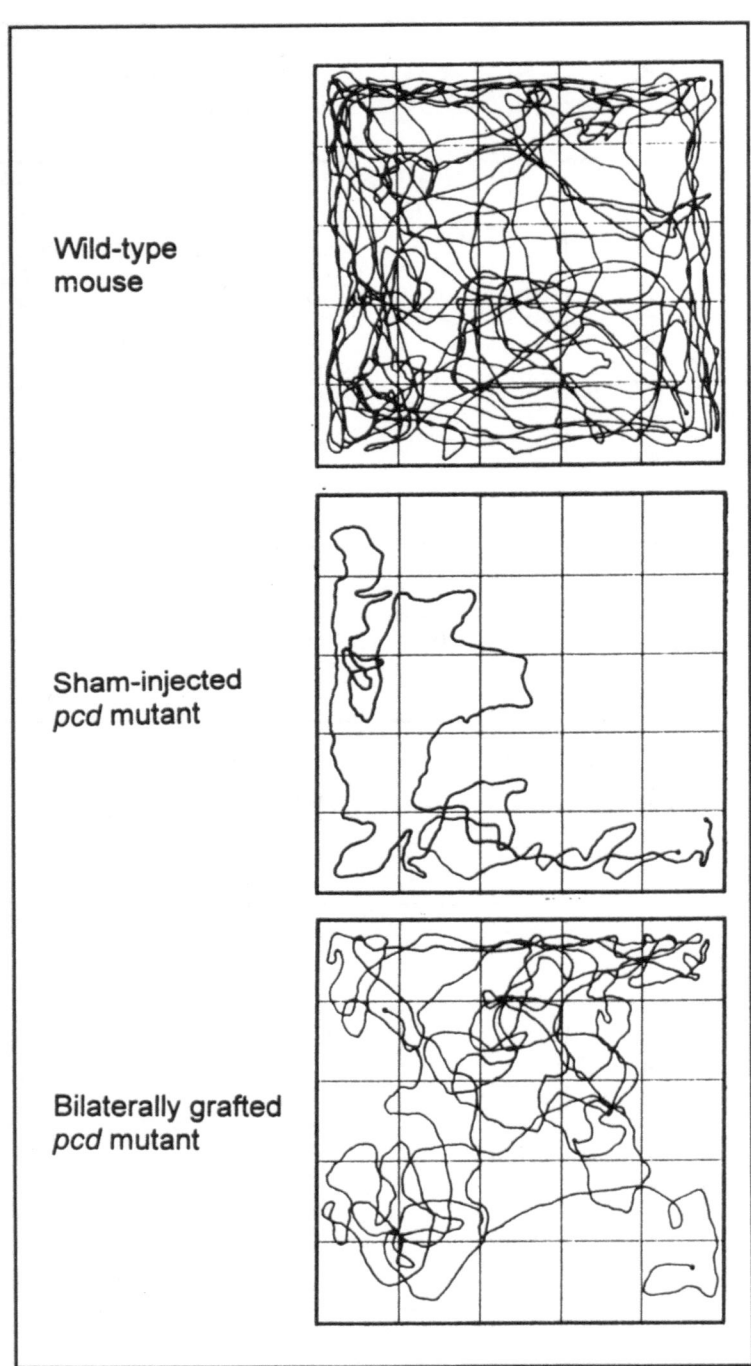

Wild-type mouse

Sham-injected
***pcd* mutant**

Bilaterally grafted
***pcd* mutant**

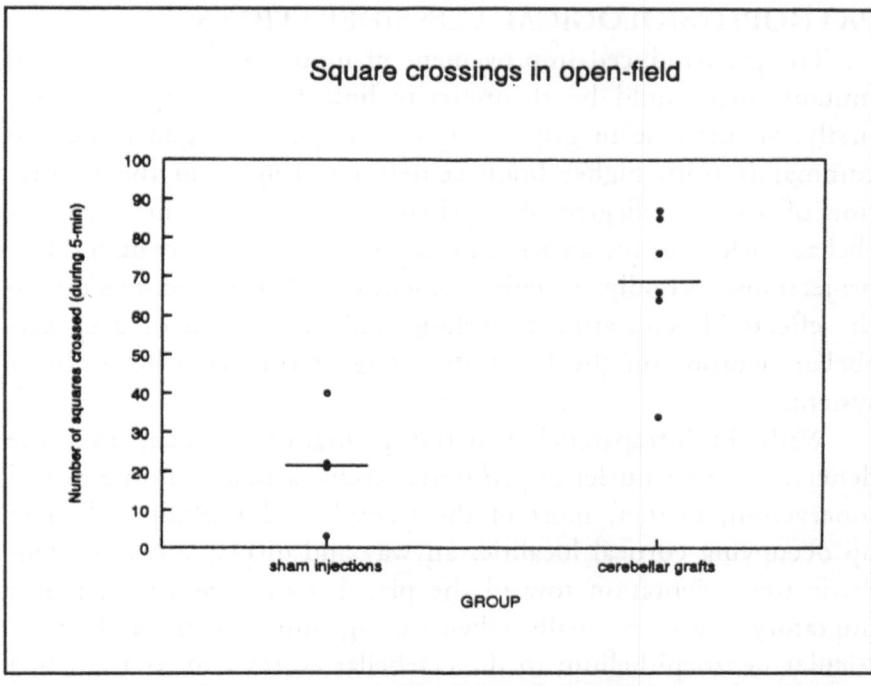

Fig. 8.5. Square crossings in the open-field matrix by sham-operated (open circles) and graft-recipient mice (filled circles). Horizontal bars indicate group means. This test provides an index of locomotor activity. With bilateral transplants, activity levels were increased on the average by more than threefold; the group difference was statistically significant at the 0.0036 probability level. In comparison, wild-type control animals registered an average of 227 crossing events during the same amount of time. Reprinted with permission from: Triarhou LC, Zhang W, Lee W-H. Cell Transpl 1996; 5:269-277. © 1996 Elsevier Science Inc.

PATHOPHYSIOLOGICAL CONSIDERATIONS

The graft-induced improvement of motor performance in *pcd* mutant mice could be theoretically linked to two components: firstly, an increase in grip strength and muscle responsiveness to commands from higher brain centers resulting from the restoration of a certain degree of physiological activity in the deep cerebellar nuclei and the associated cerebellothalamic and dentatorubral projections; secondly, an enhancement of balance functions due to the effects of reinstating a Purkinje cell innervation of deep cerebellar neurons on the functional state of the cerebellovestibular system.

With the intraparenchymal transplantation protocol used, the denervated deep nuclei of *pcd* hosts receive a new Purkinje axonal innervation; further, most of the transplanted Purkinje cells end up occupying cortical localities anyway and display a correct dendritic tree orientation toward the pia, due to recapitulation of a migratory course normally taken during ontogeny from the ventricular neuroepithelium to the cerebellar cortex and to a crossing of trajectories that allows developing Purkinje cells to establish synaptic contacts with deep neurons on the way of the perikaryon to the surface.[11-13]

The physiological advantage of placing the grafts intraparenchymally is two-fold: donor Purkinje axons are able to innervate the host deep cerebellar neurons and then migrate stereotypically to colonize cerebellocortical areas, where they can be contacted and synaptically invested by host parallel and climbing fibers, as is the case for grafts placed into the cerebellar cortex.[14-16] In that context, one index of functional responsiveness of transplanted Purkinje cells to an afferent innervation by host parallel fibers, which utilize glutamate as their neurotransmitter,[17] is the expression of GluR2/3 immunoreactivity on their postsynaptic receptive fields.[2]

It has been estimated in the mouse that loss of up to 90% of Purkinje cells produces only minor effects on the functional capabilities of the animal, indicating that 10% of the Purkinje cell complement may sustain many normal motor skills.[18] While the grafting procedure leads to an improvement of motor activity in *pcd* mutants, the performance of recipient animals is still poorer than that of wild-type controls. Such a difference could be attributed to several factors, such as: *(i)* Purkinje cell replacement is only partial, if one considers that the mouse cerebellum normally contains about 200,000 Purkinje cells;[19] *(ii)* quantitative aspects of the cerebellar reconstruction in terms of neuronal connectivity, neurotransmitter regulation mechanisms and transmitter-receptor interactions remain largely unknown and *(iii)* the extracerebellar components of the *pcd* mutant phenotype that include degeneration of retinal photoreceptors,[20] mitral cells in the olfactory bulb[21] and thalamic neurons[22] could very well compromise overall animal performance and prevent a graft-induced integral recovery.

There are several additional cellular mechanisms worthy of consideration in the functional setting. Synaptic formation between grafted and host cells, neurotransmitter release and regulation are some of the mechanisms thought of underscoring the restoration of function in neural transplantation models.[23] Neurotransmitter release and regulation, as well as transmitter-receptor interactions are further considerations, and there is now evidence that cerebellar transplants have a partial normalizing effect on the denervation-induced supersensitivity of $GABA_A$ receptors in the deep cerebellar nuclei of *pcd* mutant mice.[24]

An interesting observation has been made in the deep nuclei and vestibular nuclei of *pcd* mice, which might have important implications for the compensatory mechanisms operating in the behavioral phenotype of this mutant. Specifically, Purkinje cell loss is associated with a massive appearance of parvalbumin-immunoreactivity in neuronal somata of the deep nuclei and the vestibular nuclei, as well as with a drastic increase in the total number of both cells and terminals that are glycine-immunopositive in the mutant deep nuclei.[25] The documented colocalization of parvalbumin with glycine in deep nuclei interneurons could reflect an enhanced inhibitory activity upon large output neurons.[25]

An issue that still remains open is related to the pathophysiological implications of the partial losses of granule cells,[26] basket cells,[27] deep nuclei neurons[28] and inferior olivary neurons[29,30] that occur secondarily to Purkinje cell death in the *pcd* mutant. Such changes could have either additive or compensatory effects on the state of the functional cerebellar output. Moreover, the persistence of the noradrenergic and serotoninergic innervations of the cerebellum and the ensuing remodeling after the degeneration of their target neurons[31-33] could very likely influence the outcome of cerebellar activity after transplantation.

A PROTOCOL FOR FURTHER STUDIES TO ASSESS THE EXTENT AND LIMITS OF FUNCTIONAL RECOVERY

To precisely define the extent and limitations of Purkinje cell transplantation into the cerebellum of *pcd* mutants in order to counteract the disabilities induced by the genetic mutation, a battery of behavioral tasks can be used in the future in determining the recovery of behavioral responses after unilateral or bilateral grafting.

In humans, midline-vermal cerebellar dysfunction results principally in a characteristic stance and unsteady gait with legs thrust widely apart, arms extended in compensatory balance, and cautious walking accomplished in short steps; lateral-hemispheric cerebellar abnormalities produce ipsilateral disorders of posture and movement, including hypotonia, limb incoordination, an irregular swaying gait and a tendency to drift toward the side of the lesion.[34,35] Since the cerebellar atrophy in *pcd* mutants virtually involves the entire cerebellum, their behavioral phenotype includes signs related to both vermal and bilateral hemispheric lesions. Naïve *pcd* mice manifest an ataxic syndrome beginning at 3-4 weeks of age that includes a lowered, widened stance, with an inability to sustain the abdomen in a raised relief, and the hind-limbs in an abducted and hyperextended position.

By E12, which is the donor tissue age at the time of graft harvesting, only Purkinje cells have undergone their final mitotic division in the primordial cerebellar cortex, while the remaining cortical interneurons have not been generated yet; gliogenesis occurs even later in development.[36,37] Therefore, there is a very good chance that a large portion of the suspension cells are Purkinje cells. Based on the behavioral data already obtained, it appears that the number of surviving Purkinje cells suffices in bringing about some functional recovery.

ASSESSMENT OF FUNCTIONAL RECOVERY
AFTER UNILATERAL TRANSPLANTATION

On walking, patients with unilateral hemispheric lesions stagger and progressively deviate to the affected side.[34] This can be demonstrated by asking the patient to walk around a chair. As the patients rotate toward the affected side, they tend to fall into the chair; rotating toward the normal side, they move away from the chair in a spiral.

There is a well-established model of rotational asymmetry in rodents with unilateral lesions of specific brain pathways.[38] Rotometers have been used previously to assess graft-induced functional recovery in a dopaminergic deficit mouse model.[39] The animals are placed inside a cylindrical chamber and connected through a Velcro® vest to a mechanical-electrical transducer through a wire string. Turning behavior is monitored as total number of full body turns in each direction for adjustable time intervals electronically.

Based on the rotational pattern in unilateral human cerebellar lesions described above, one may predict that *pcd* mutants with unilateral grafts of cerebellar cell suspensions into the deep cerebellar nuclei of one hemisphere will rotate toward the nongrafted side, which relative to the transplanted side will appear as the affected hemisphere.

ASSESSMENT OF FUNCTIONAL RECOVERY
AFTER BILATERAL TRANSPLANTATION

The description of behavioral tests that follows involves motoric tasks in which the optimal performance is based on bilateral replenishment of Purkinje cells in the atrophic host cerebellum. Various tests could be carried out to determine the extent and limits of graft-induced functional recovery.

Activity monitoring

Activity levels can be measured accurately by means of a stainless steel activity wheel 14" in diameter, that also includes a resting cage and revolution counter (Lafayette Instruments, Lafayette, IN). The force required to spin the wheel can be custom calibrated and adjusted. The resting cage measures 10" × 6" × 5". A micro-switch connects the wheel to an electronic impulse counter to record the number of revolutions that the animal has run.

Ergometric activity

A small monitor measures animal movements by means of a platform accelerometer (Coulbourn Instruments, Lehigh Valley, PA). Data output consists of force-displacement/time integral count, and is proportional to the ergometric activity of the mice. The unit measures small nonambulatory movements, including grooming and tremors, producing a proportionally slow time integral count rate. Photobeam activity monitors produce data based on fixed spatial movements as the animal moves about and will not register grooming, tremor or other movements between the beams. Photobeam activity monitors are used to measure where the animal goes; the accelerometric movement monitor measures how much effort the animal expends. The control/data monitor unit permits automatic session timing, data output scaling and movement separation into two counters. The dimensions of the accelerometer are $9" \times 9" \times 2\frac{1}{2}"$, and sensing is based on displacement-force against mass and a small component of g.

Strength measurement

Animal deficiency in strength can be measured objectively through a "grip strength meter" designed for mice (Columbus Instruments International, Columbus, OH). The forelimb grip bar consists of a rectangular wire mesh measuring $3" \times 2"$ parallel to the plane of the animal's body, and the hindlimb grip bar is a $5"$ rectangular grid, angled $45°$ to facilitate hindlimb grasping. The animal platform is constructed of aluminum and measures $31" \times 11" \times 13"$. Four sensors located inside a $9\frac{1}{2}"$ unit are connected to an analog dial readout. Force units can be displayed in grams (0-1000 g), pounds (0-2 lb) or newtons (1-10 N) with an accuracy of $\pm0.15\%$ of the full scale reading.

Capacitance field with electromagnetic detection

A body motion monitor can measure animal activity based on the principle of capacitance change (Columbus Instruments International, Columbus, OH). A standard plastic cage is used for animal housing during the experimental session. An active electrode is located beneath the plastic top of the monitor and a flat metal lid is located on top of the animal cage and grounded to the monitor. Animal movement between the two electrodes produces changes in electromagnetic field capacitance, and pulses are sent to the counters. Measurement results are relative to animal size. Results are presented as number of animal movements. Sensitivity is adjusted by panel dials. Other than sensitivity adjustments, there is no necessity for tuning. Sensitivity is stable and independent of water, urine or feces in the cage. The monitor is equipped with two precise sensitivity thresholds and two electronic counters. It allows the separation of small vertical movements associated with running from large vertical movements such as rearing and jumping. Results are related to composed activity of the animals and to the number of movements, but not to the distance animals cover. In the "two-sensitivities" mode, the counter with the higher sensitivity registers both smaller and larger movements, while the counter with the lower sensitivity registers larger signals only. In "D" mode, the counter with the higher sensitivity registers small signals only, while a second counter registers larger signals.

A model for nonlinear systems analysis of capacitance field based monitoring

The analysis of capacitance field based data can be based on ideas from nonlinear dynamical systems theory, where a universality is present, expressed as a unique form in the log-log plot of spectral frequencies demonstrating $1/f$ noise.[40] The occurrence of the $1/f$ noise phenomenon[41] is attributed to the richness of the possible animal movements and is related to multiplicative and broad log-normal distributions.[42-44] The $1/f$ noise or flicker noise is a common phenomenon describing fluctuations in many areas of physics,[42,45] and it may be due to a rich random statistical ensemble with atypical configurations dominating over the usual mean values. In all, capacitance field-based behavior has a power-law spectrum of fluctuations characterized by Fourier transforms decaying as $1/f$, similar to the complex dynamics encoded by universal computing machines known as cellular automata.[45] These methods favor a nonlinear systems-based approach in behavioral analysis as a fruitful alternative to the widely used stochastic methods.[46]

THE CEREBELLUM AND HIGHER BRAIN FUNCTIONS

Apart from movement control, the cerebellum is believed to participate in certain higher cognitive functions, including spatial working memory,[47] motor program elaboration,[48] habituation exploration behavior,[49] memory formation for eyeblink conditioning[50-52] and computational attempts.[53] Behavioral studies in intact and cerebellectomized Lurcher mutant mice indicate that the deep cerebellar nuclei are sufficient for motor learning, provided that the task is not too difficult, but that the cerebellar cortex is required when a more difficult task is involved.[54]

In addition to defining the extent and limits of the recovery of motor responses after cerebellar transplantation in the *pcd* ataxic mouse model, it might be worthwhile to explore possible effects of the grafts on more complex behaviors, such as eyeblink conditioning,[52] distal-cue spatial navigation and visual guidance performance.[55]

REFERENCES

1. Triarhou LC, Zhang W, Lee W-H. Graft-induced restoration of function in hereditary cerebellar ataxia. Neuroreport 1995; 6:1827-1832.
2. Triarhou LC, Zhang W, Lee W-H. Amelioration of the behavioral phenotype in genetically ataxic mice through bilateral intracerebellar grafting of fetal Purkinje cells. Cell Transpl 1996; 5:269-277.
3. Zhang W, Lee W-H, Triarhou LC. Grafted cerebellar cells in a mouse model of hereditary ataxia express IGF-I system genes and partially restore behavioral function. Nature Med 1996; 2:65-71.
4. Triarhou LC. The cerebellar model of neural grafting: Structural integration and functional recovery. Brain Res Bulletin 1996; 39:127-138.
5. Triarhou LC, Low WC, Ghetti B. Intraparenchymal grafting of cerebellar cell suspensions to the deep cerebellar nuclei of *pcd* mutant mice, with particular emphasis on re-establishment of a Purkinje cell cortico-nuclear projection. Anat Embryol (Berl) 1992; 185:409-420.
6. Bureš J, Burešová O, Huston J. Techniques and Basic Experiments for the Study of Brain and Behavior. Amsterdam: Elsevier Scientific, 1976:37-89.
7. Cory-Slechta DA. Behavioral measures of neurotoxicity. Neurotoxicology 1989; 10:271-296.
8. Ishimatsu S, Igisu H, Tanaka I. A simple apparatus to measure weakness and ataxia in the mouse. Neurotoxicology 1990; 11:719.

9. Jones BJ, Roberts DJ. The quantitative measurement of motor in-coordination in naïve mice using an accelerating rotarod. J Pharm Pharmacol 1968; 20:302-304.

10. Pellegrino LJ, Altman J. Effects of differential interference with postnatal cerebellar neurogenesis on motor performance, activity level, and maze learning of rats: A developmental study. J Comp Physiol Psychol 1979; 93:1-33.

11. Altman J, Bayer SA. Embryonic development of the rat cerebellum. II. Translocation and regional distribution of the deep neurons. J Comp Neurol 1985; 231:27-41.

12. Altman J, Bayer SA. Embryonic development of the rat cerebellum. III. Regional differences in the time of origin, migration, and settling of Purkinje cells. J Comp Neurol 1985; 231:42-65.

13. Yuasa S, Kawamura K, Ono K et al. Development and migration of Purkinje cells in the mouse cerebellar primordium. Anat Embryol (Berl) 1991; 184:195-212.

14. Gardette R, Alvarado-Mallart RM, Crepel F et al. Electrophysiological demonstration of a synaptic integration of transplanted Purkinje cells into the cerebellum of the adult Purkinje cell degeneration mutant mouse. Neuroscience 1988; 24:777-789.

15. Sotelo C, Alvarado-Mallart RM. Growth and differentiation of cerebellar suspensions transplanted into the adult cerebellum of mice with heredodegenerative ataxia. Proc Natl Acad Sci USA 1986; 83:1135-1139.

16. Sotelo C, Alvarado-Mallart RM. Reconstruction of the defective cerebellar circuitry in adult Purkinje cell degeneration mutant mice by Purkinje cell replacement through transplantation of solid embryonic implants. Neuroscience 1987; 20:1-22.

17. Hudson DB, Valcana T, Bean G et al. Glutamic acid: a strong candidate as the neurotransmitter of the cerebellar granule cell. Neurochem Res 1976; 1:73-81.

18. Wetts R, Moran T, Oster-Granite M et al. Effect of Purkinje cell loss on complex motor behavior. Soc Neurosci Abstr 1985; 11:1037.

19. Caddy KWT, Biscoe TJ. Structural and quantitative studies in the normal C3H and Lurcher mutant mouse. Phil Trans Roy Soc Lond (Biol) 1979; 287:167-201.

20. Mullen RJ, LaVail MM. Two new types of retinal degeneration in cerebellar mutant mice. Nature (Lond) 1975; 258:528-530.

21. Greer CA, Shepherd GM. Mitral cell degeneration and sensory function in the neurological mutant mouse Purkinje cell degeneration. Brain Res 1982; 235:156-161.

22. O'Gorman S, Sidman RL. Degeneration of thalamic neurons in 'Purkinje cell degeneration' mutant mice. I. Distribution of neuron loss. J Comp Neurol 1985; 234:277-297.

23. Björklund A, Lindvall O, Isacson O et al. Mechanisms of action of intracerebral neural implants: Studies on nigral and striatal grafts to the lesioned striatum. Trends Neurosci 1987; 10:509-516.

24. Mitsacos A, Stasi K, Kouvelas ED et al. Functional integration of transplanted Purkinje cells into the atrophic cerebellum: II. Inhibitory amino acid receptors and efferent innervation. Abstr Am Soc Neural Transpl 1996; 3:50.

25. Bäurle J, Grüsser-Cornehls U. Increased somatal *parvalbumin* and glycine immunoreactivity in the cerebellar targets of Purkinje cell degeneration mutants. Soc Neurosci Abstr 1995; 21:2081.

26. Triarhou LC, Norton J, Alyea C et al. A quantitative study of the granule cells in the Purkinje cell degeneration *(pcd)* mutant. Ann Neurol 1985; 18:146.

27. Ghetti B, Triarhou LC. The Purkinje cell degeneration mutant: A model to study the consequences of neuronal degeneration. In: Plaitakis A, ed. Cerebellar Degenerations: Clinical Neurobiology. Boston: Kluwer Academic Publishers, 1992:159-181.

28. Triarhou LC, Norton J, Ghetti B. Anterograde transsynaptic degeneration in the deep cerebellar nuclei of Purkinje cell degeneration *(pcd)* mutant mice. Exp Brain Res 1987; 66:577-588.

29. Ghetti B, Norton J, Triarhou LC. Nerve cell atrophy and loss in the inferior olivary complex of "Purkinje cell degeneration" mutant mice. J Comp Neurol 1987; 260:409-422.

30. Triarhou LC, Ghetti B. Stabilisation of neurone number in the inferior olivary complex of aged 'Purkinje cell degeneration' mutant mice. Acta Neuropathol (Berl) 1991; 81:597-602.

31. Triarhou LC, Ghetti B. Monoaminergic nerve terminals in the cerebellar cortex of Purkinje cell degeneration mutant mice: Fine structural integrity and modification of cellular environs following loss of Purkinje and granule cells. Neuroscience 1986; 18:795-807.

32. Triarhou LC, Ghetti B. Serotonin immunoreactivity in the cerebellum of two neurological mutant mice and the corresponding wild-type genetic stocks. J Chem Neuroanat 1991; 4:421-428.

33. Triarhou LC, Low WC, Ghetti B. Serotonin fiber innervation of cerebellar cell suspensions intraparenchymally grafted to the cerebellum of *pcd* mutant mice. Neurochem Res 1992; 17:475-482.

34. Gilman S. Gait disorders. In: Rowland LP, ed. Merritt's Textbook of Neurology. Philadelphia: Lea & Febiger, 1989:54-56.

35. Plum F. Ataxia and related gait disorders. In: Wyngaarden JB, Smith LH Jr, Bennett JC, eds. Cecil Textbook of Medicine. 19th ed. Philadelphia: Saunders, 1992:2113-2115.

36. Miale IL, Sidman RL. An autoradiographic analysis of histogenesis in the mouse cerebellum. Exp Neurol 1961; 4:277-296.

37. Altman J. Morphological development of the rat cerebellum and some of its mechanisms. Exp Brain Res [Suppl] 1982; 6:8-49.

38. Ungerstedt U, Arbuthnott GW. Quantitative recording of rotational behavior in rats after 6-hydroxydopamine lesions of the nigrostriatal dopamine system. Brain Res 1970; 24:485-493.

39. Witt TC, Triarhou LC. Transplantation of mesencephalic cell suspensions from wild-type and heterozygous weaver mice into the denervated striatum: Assessing the role of graft-derived dopaminergic dendrites in the recovery of function. Cell Transpl 1995; 4:323-333.

40. Kafetzopoulos E, Gouskos S, Evangelou SN. The fractal geometry of behaviour. Proceedings of IBRO Workshop on Mechanisms of Neuronal Plasticity, University of Patras, 1992.

41. Kobayashi M, Musha T. $1/f$ fluctuation of heartbeat period. IEEE Trans Biomed Eng 1982; 29:456-457.

42. Shlesinger MF. Fractal time and $1/f$ noise in complex systems. Ann NY Acad Sci 1987; 504:214-228.

43. Goldberger AL, West BJ. Chaos and order in the human body. MD Computing 1992; 9:25-34.

44. McKenna TM, McMullen TA, Shlesinger MF. The brain as a dynamic physical system. Neuroscience 1994; 60:587-605.

45. Bak P, Chen K, Creutz M. Self-organized criticality in the 'Game of Life'. Nature (Lond) 1989; 342:780-782.

46. Pinsker HM, Willis WD, eds. Information Processing in the Nervous System. New York: Raven Press, 1980.

47. Middleton FA, Strick PL. Anatomical evidence for cerebellar and basal ganglia involvement in higher cognitive function. Science 1994; 266:458-461.

48. Dahhaoui M, Lannou J, Stelz T et al. Role of the cerebellum in spatial orientation in the rat. Behav Neural Biol 1992; 58:180-189.

49. Dahhaoui M, Caston J, Lannou J et al. Role of the cerebellum in habituation exploration behavior in the rat. Physiol Behav 1992; 52:339-344.

50. Krupa DJ, Thompson JK, Thompson RF. Localization of a memory trace in the mammalian brain. Science 1993; 260:989-991.

51. Aiba A, Kano M, Chen C et al. Deficient cerebellar long-term depression and impaired motor learning in mGluR1 mutant mice. Cell 1994; 79:377-388.

52. Freeman JH Jr, Barone S Jr, Stanton ME. Disruption of cerebellar maturation by an antimitotic agent impairs the ontogeny of eyeblink conditioning in rats. J Neurosci 1995; 15:7301-7314.

53. Kim SG, Ugurbil K, Strick PL. Activation of a cerebellar output nucleus during cognitive processing. Science 1994; 265:949-951.

54. Caston J, Vasseur F, Stelz T et al. Differential roles of cerebellar

cortex and deep cerebellar nuclei in the learning of the equilibrium behavior: Studies in intact and cerebellectomized Lurcher mutant mice. Dev Brain Res 1995; 86:311-316.

55. Goodlett CR, Hamre KM, West JR. Dissociation of spatial navigation and visual guidance performance in Purkinje cell degeneration *(pcd)* mutant mice. Behav Brain Res 1992; 47:129-141.

CHAPTER 9

CLINICAL POTENTIAL

CLINICAL NEURAL TRANSPLANTATION TRIALS IN HUMAN NEURODEGENERATIVE CONDITIONS

Clinical trials with fetal mesencephalic grafts into the caudate nucleus or putamen have been reported in parkinsonian patients in medical centers of several countries, including Sweden,[1-16] United Kingdom,[17-28] Mexico,[29,30] United States,[31-47] Cuba,[48] Russia,[49] Czech Republic,[50] Slovakia,[51] Canada,[52] Spain,[53,54] China,[55] Poland[56] and France.[57] Such trials have been prompted by encouraging results from extensive experimental studies in rodent and primate models. Evidence for graft survival[4,12,13,16,26,28,35,42,44,57] and functional improvement of clinical signs[1-29,31-57] has been presented in many of those studies. Reported variations in the outcome of the procedure might relate *inter alia* to technique and site of grafting, age and method of preparation of donor tissue, stage of advancement of the disease in the host and pharmacological scheme of the patients prior to the operation.

The rationale for using fetal striatal cell grafts into the striatum of patients with Huntington's chorea has been presented[58-61] and initial clinical trials have also been attempted.[62,63]

In the United States, the Registry Committee of the American Society for Neural Transplantation collects basic demographic, morbidity and mortality data from clinical neural transplantation studies and carries out extensive efficacy evaluations.[64] In the European Union, the Network of European CNS Transplantation and Restoration has been founded to carry out a concerted effort for the development of efficient, reliable, safe and ethically acceptable transplantation therapies for neurodegenerative diseases.[65,66]

OPEN ISSUES AND FUTURE DIRECTIONS
OF CEREBELLAR TRANSPLANTATION

Purkinje cell loss is the histopathological hallmark of a host of hereditary and sporadic cerebellar atrophies, as well as neurotoxic, anoxic-ischemic and paraneoplastic encephalopathies. New discoveries on the etiology and pathogenesis of cerebellar disorders may lead to a prevention or arrest of the pathological process. One way of counteracting neurological disease in the future may be through early detection my means of molecular genetic or neuroimaging methods and subsequent trophic or pharmacological supplementation in order to block further progression. Nonetheless, for all practical purposes, there will most likely always be conditions with irreversible loss of cerebellar neurons where replacement through transplantation of homologous tissue might in theory be the only realistic option.

Undoubtedly, neural transplantation is a controversial form of cell replacement therapy. However, the encouraging results obtained from experimental studies in animal models and from the initial clinical trials in neurodegenerative disorders such as Parkinson's disease and Huntington's chorea reinforce the idea that basic research on cerebellar grafting with its implications for the treatment of the human ataxias has its merit. As a tool for biomedical research, neural transplantation will continue to be used to study aspects of cerebellar development and plasticity and to provide the necessary background for a future clinical application. The field can substantially benefit from the interaction of transplantation neurobiologists, clinical neuroscientists and basic cell biologists.

The anatomical, pharmacological and behavioral studies conducted so far in genetically ataxic mice with cerebellar grafts have: *(i)* monitored the migratory processes of transplanted Purkinje cells; *(ii)* described the re-establishment of synaptic connectivity; *(iii)* indicated a normalizing effect of the grafts on the denervation-induced receptor supersensitivity and *(iv)* demonstrated a partial improvement of the behavioral phenotype.

The detection of demonstrable effects after transplantation is only an initial step and by no means the end. Many questions remain unanswered, and additional studies are necessary to further decipher numerical aspects of neuronal survival and replacement, neurotransmitter regulation and receptor interactions, quantitative aspects of the integration with the host brain, as well as the extent and limits of functional recovery. Due to the precise nature of its construction, the cerebellum is a fairly difficult anatomical system to restructure in its entirety; therefore, one has to give priority to certain components of circuit reinstatement over others, depending on the particular experimental aim.

The importance of the surgical implantation approach in relationship to the migratory properties of the grafts and the geometry of the host cerebellar nuclei needs to be emphasized. The way of graft placement is a crucial point in neural transplantation research, and important advances in the field have been made through modifications of grafting techniques in animal studies. Tissue placement may dictate the functional effects of the grafts.

Another important topic is that of cografted cerebellar neurons in addition to Purkinje cells. By grafting the cerebellar anlage one cannot automatically assume that it only contains Purkinje cells; other neurons in the grafts may also utilize γ-aminobutyric acid (GABA), and their functional contribution needs to be dissected out. Perhaps in the years ahead one may be in a position to obtain clean preparations of Purkinje cells for grafting purposes.

Finally, trophic factor supplementation of the grafts, especially at the time of physical separation from their normal environs during the dissection from the fetal brain, is a topic that needs to be investigated.

In the Parkinson's disease model, the growth of human fetal mesencephalic neurons after transplantation was first monitored in human-to-rat grafting experiments.[67,68] Regarding the cerebellum, the growth properties of human Purkinje cells are being studied after transplantation into the cerebellum of nude (athymic) mice, where they are not rejected due to the deficiency of these mice in cellular immunity.[69]

The cerebellar transplantation studies in rodents have provided encouraging results that underscore the potential of neural transplantation for restoring cerebellar function. There are still gaps in knowledge with regard to the pathological, structural and functional aspects of the interaction between the grafts and the host cerebellum, which need to be bridged through experimental studies before contemplating the feasibility of applying neural transplantation to patients with cerebellar atrophies.

The chief objective of neurobiological research in the field of cerebellar transplantation is to explore an innovative experimental treatment for a host of incurable and debilitating neurological disorders that affect the cerebellum and for which there is no known cure at present. The prospect of hope for patients with certain forms of cerebellar ataxia is exciting. As always, one will have to find the golden section between the potential alleviation of neurological symptoms and the medical risks involved, giving full consideration to the results from basic studies in transplantation neurobiology, and being guided by the Hippocratic principles of "speaking the past, knowing the present, foretelling the future; studying these; and on diseases, making a habit of two things, doing good or at least doing no harm".[70]

CONCLUDING REMARKS

Neural transplantation has been successfully applied to replace degenerated neurons in several anatomical systems experimentally[71,72] and in clinical trials in patients with Parkinson's[1-12,15-28,31-57] and Huntington's disease.[63] A distinction has been made between 'global' or 'paracrine' systems (such as, e.g., the mesostriatal dopamine projection), in which local release of neurotransmitter may suffice for recovery, and 'point-to-point' systems (such as the cerebellum), where a precise re-establishment of the missing circuitry is deemed necessary,[73,74] although synaptic formation is considered as one of the mechanisms underlying the recovery of function in global systems as well.[75]

The functional effects of neural transplants on motor performance in 'point-to-point' systems had long remained an open question.[76] The behavioral findings reviewed in chapter 8 provide evidence for motor enhancement in an ataxic mouse model after intracerebellar transplantation of fetal Purkinje neurons, thus lending credence to the thesis that neural grafting is a viable approach in restoring function not only in diffuse 'paracrine' systems, but in neural systems characterized by 'point-to-point' synaptic connectivity as well, and underscoring the clinical potential for future cerebellar neuron implantation in counteracting certain forms of the human cerebellar ataxias. At present, however, the application of cerebellar neuron implantation in humans with cerebellar ataxia[77,78] still seems premature, as many of the pathological and biochemical mechanisms in the interaction between grafted tissue and the host brain need to be further elucidated in extensive experimental studies, and great caution as well as strict criteria must be used in contemplating the theoretical feasibility of a possible application of cerebellar transplants in humans.

REFERENCES

1. Lindvall O, Rehncrona S, Gustavii B et al. Fetal dopamine-rich mesencephalic grafts in Parkinson's disease. Lancet 1988; ii: 1483-1484.
2. Lindvall O, Rehncrona S, Brundin P et al. Human fetal dopamine neurons grafted into the striatum in two patients with severe Parkinson's disease: A detailed account of methodology and a 6-month follow-up. Arch Neurol 1989; 46:615-631.
3. Lindvall O. Transplantation into the human brain: Present status and future possibilities. J Neurol Neurosurg Psychiat 1989; Suppl:39-54.
4. Lindvall O, Brundin P, Widner H et al. Grafts of fetal dopamine neurons survive and improve motor function in Parkinson's disease. Science 1990; 247:574-577.
5. Brundin P, Odin P, Widner H. Promising new results with transplantation of nerve cells to the brain in Parkinson disease. Lakartidningen 1990; 87:3761-3763.
6. Brundin P, Björklund A, Lindvall O. Practical aspects of the use of human fetal brain tissue for intracerebral grafting. Prog Brain Res 1990; 82:707-714.
7. Lindvall O, Rehncrona S, Brundin P et al. Neural transplantation in Parkinson's disease: The Swedish experience. Prog Brain Res 1990; 82:729-734.
8. Lindvall O. Prospects of transplantation in human neurodegenerative diseases. Trends Neurosci 1991; 14:376-384.
9. Lindvall O, Björklund A, Widner H, eds. Intracerebral Transplantation in Movement Disorders: Experimental Basis and Clinical Experiences. Amsterdam: Elsevier, 1991.
10. Widner H, Brundin P, Rehncrona S et al. Transplanted allogeneic fetal dopamine neurons survive and improve motor function in idiopathic Parkinson's disease. Transpl Proc 1991; 23:793-795.
11. Lindvall O. Transplants in Parkinson's disease. Eur Neurol 1991; 31[Suppl 1]:17-27.

12. Lindvall O, Widner H, Rehncrona S et al. Transplantation of fetal dopamine neurons in Parkinson's disease: One-year clinical and neurophysiological observations in two patients with putaminal implants. Ann Neurol 1992; 31:155-165.
13. Widner H, Tetrud J, Rehncrona S et al. Bilateral fetal mesencephalic grafting in two patients with parkinsonism induced by 1-methyl-4-phenyl-1,2,3,6-tetrahydropyridine (MPTP). N Engl J Med 1992; 327:1556-1563.
14. Widner H, Tetrud J, Rehncrona S et al. Fifteen months' follow-up on bilateral embryonic mesencephalic grafts in two cases of severe MPTP-induced parkinsonism. Adv Neurol 1993; 60:729-733.
15. Widner H, Rehncrona S. Transplantation and surgical treatment of parkinsonian syndromes. Curr Opin Neurol Neurosurg 1993; 6:344-349.
16. Lindvall O, Sawle G, Widner H et al. Evidence for long-term survival and function of dopaminergic grafts in progressive Parkinson's disease. Ann Neurol 1994; 35:172-180.
17. Hitchcock ER, Clough C, Hughes R et al. Embryos and Parkinson's disease. Lancet 1988; i:1274.
18. Hitchcock ER, Kenny BG, Clough CG et al. Stereotactic implantation of fetal mesencephalon. Stereotact Funct Neurosurg 1990; 54/55:282-289.
19. Quinn NP. The clinical application of cell grafting techniques in patients with Parkinson's disease. Prog Brain Res 1990; 82:619-625.
20. Hitchcock ER, Kenny BG, Clough CG et al. Stereotactic implantation of foetal mesencephalon (STIM): The UK experience. Prog Brain Res 1990; 82:723-728.
21. Henderson BTH, Kenny BG, Hitchcock ER et al. A comparative evaluation of clinical rating scales and quantitative measurements in assessment pre and post striatal implantation of human foetal mesencephalon in Parkinson's disease. Acta Neurochir 1991; Suppl 52:48-50.
22. Henderson BT, Clough CG, Hughes RC et al. Implantation of human fetal ventral mesencephalon to the right caudate nucleus in advanced Parkinson's disease. Arch Neurol 1991; 48:822-827.
23. Hitchcock ER, Kenny BG, Henderson BTH et al. A series of experimental surgery for advanced Parkinson's disease by foetal mesencephalic transplantation. Acta Neurochir 1991; Suppl 52:54-57.
24. Hitchcock ER. Neural implants and recovery of function: Human work. Adv Exp Med Biol 1992; 325:67-78.
25. Sinden JD, Patel SN, Hodges H. Neural transplantation: Problems and prospects for therapeutic application. Curr Opin Neurol Neurosurg 1992; 5:902-908.
26. Sawle GV, Bloomfield PM, Björklund A et al. Transplantation of

fetal dopamine neurons in Parkinson's disease: PET [^{18}F]6-L-fluorodopa studies in two patients with putaminal implants. Ann Neurol 1992; 31:166-173.

27. Henderson B, Good PA, Hitchcock ER et al. Visual evoked cortical responses and electroretinograms following implantation of human fetal mesencephalon to the right caudate nucleus in Parkinson's disease. J Neurol Sci 1992; 107:183-190.

28. Sawle GV, Myers R. The role of positron emission tomography in the assessment of human neurotransplantation. Trends Neurosci 1993; 16:172-176.

29. Madrazo I, León V, Torres C et al. Transplantation of fetal substantia nigra and adrenal medulla to the caudate nucleus in two patients with Parkinson's disease. N Engl J Med 1988; 318:51.

30. Madrazo I, Franco-Bourland R, Ostrosky-Solis F et al. Neural transplantation (auto-adrenal, fetal nigral and fetal adrenal) in Parkinson's disease: The Mexican experience. Prog Brain Res 1990; 82:593-602.

31. Freed CR, Breeze RE, Rosenberg NL et al. Transplantation of human fetal dopamine cells for Parkinson's disease: Results at 1 year. Arch Neurol 1990; 47:505-512.

32. Freed CR, Breeze RE, Rosenberg NL et al. Therapeutic effects of human fetal dopamine cells transplanted in a patient with Parkinson's disease. Prog Brain Res 1990; 82:715-721.

33. Fiandaca MS. Brain grafting for Parkinson's disease. Transplantation 1991; 51:549-556.

34. Spencer DD, Robbins RJ, Naftolin F et al. Unilateral transplantation of human fetal mesencephalic tissue into the caudate nucleus of patients with Parkinson's disease. N Engl J Med 1992; 327:1541-1548.

35. Freed CR, Breeze RE, Rosenberg NL et al. Survival of implanted fetal dopamine cells and neurologic improvement 12 to 46 months after transplantation for Parkinson's disease. N Engl J Med 1992; 327:1549-1555.

36. Bakay RAE. Central nervous system grafting: Animal and clinical results. Stereotact Funct Neurosurg 1992; 58:67-78.

37. Thompson L. Fetal transplants show promise. Science 1992; 257:868-870.

38. Langston JW, Widner H, Goetz CG et al. Core assessment program for intracerebral transplantation (CAPIT). Movt Dis 1992; 7:2-13.

39. Goetz CG, De Long MR, Penn RD et al. Neurosurgical horizons in Parkinson's disease. Neurology 1993; 43:1-7.

40. Freed CR, Breeze RE, Rosenberg NL et al. Embryonic dopamine cell implants as a treatment for the second phase of Parkinson's disease: Replacing failed nerve terminals. Adv Neurol 1993; 60:721-728.

41. Redmond DE Jr, Robbins RJ, Naftolin F et al. Cellular replacement of dopamine deficit in Parkinson's disease using human fetal mesencephalic tissue: Preliminary results in four patients. Res Publ Assoc Res Nerv Ment Dis 1993; 71:325-359.

42. Rauch RA, Markham CH, Rand RW et al. MR imaging findings after transplant surgery for Parkinson disease. J Magn Reson Imag 1994; 4:19-24.

43. Freeman TB, Olanow CW, Hauser RA et al. Bilateral fetal nigral transplantation into the postcommissural putamen in Parkinson's disease. Ann Neurol 1995; 38:379-388.

44. Kordower JH, Freeman TB, Snow BJ et al. Neuropathological evidence of graft survival and striatal reinnervation after the transplantation of fetal mesencephalic tissue in a patient with Parkinson's disease. N Engl J Med 1995; 332:1118-1124.

45. Price LH, Spencer DD, Marek KL et al. Psychiatric status after human fetal mesencephalic tissue transplantation in Parkinson's disease. Biol Psychiat 1995; 38:498-505.

46. Olanow CW, Kordower JH, Freeman TB. Fetal nigral transplantation as a therapy for Parkinson's disease. Trends Neurosci 1996; 19:102-109.

47. Kopyov OV, Jacques DS, Lieberman A et al. Clinical study of fetal mesencephalic intracerebral transplants for the treatment of Parkinson's disease. Cell Transpl 1996; 5:327-337.

48. Molina H, Quiñones R, Alvarez L et al. Transplantation of human fetal mesencephalic tissue in caudate nucleus as treatment for Parkinson's disease: The Cuban experience. Restor Neurol 1991; 4:99-110.

49. Bekhtereva NP, Gilerovich EG, Gurchin FA et al. Transplantation of embryonal nerve tissues in the treatment of Parkinson disease. Z Nevropatol Psikhiat SS Korsakova 1990; 90:10-13.

50. Subrt O, Tichy M, Vladyka V et al. Grafting of fetal dopamine neurons in Parkinson's disease: The Czech experience with severe akinetic patients. Acta Neurochir 1991; Suppl 52:51-53.

51. Marsala J, Zigova T, Badonic T et al. Neurotransplantation, critical analysis and perspectives. Bratislav Lekar List 1992; 93:111-122.

52. Jones D. Halifax hospital first in Canada to proceed with controversial fetal-tissue transplant. Can Med Assoc J 1992; 146:389-391.

53. Lopez-Lozano JJ, Brera B, Bravo G et al. Neural transplants in Parkinson's disease. Transpl Proc 1993; 25:1005-1011.

54. Lopez-Lozano JJ, Bravo G, Brera B et al. Long-term follow-up in 10 Parkinson's disease patients subjected to fetal brain grafting into a cavity in the caudate nucleus: The Clinica Puerta de Hierro experience. Transpl Proc 1995; 27:1395-1400.

55. Iacono RP, Tang ZS, Mazziotta JC et al. Bilateral fetal grafts for

Parkinson's disease: 22 months' results. Stereotact Funct Neurosurg 1992; 58:84-87.

56. Zabek M, Mazurowski W, Dymecki J et al. Transplantation of fetal dopaminergic cells in Parkinson disease. Neurol Neurochir Polsk 1992; Suppl 1:13-19.

57. Remy P, Samson Y, Hantraye P et al. Clinical correlates of [18F]fluorodopa uptake in five grafted parkinsonian patients. Ann Neurol 1995; 38:580-588.

58. Sanberg PR, Wictorin K, Isacson O. Cell Transplantation for Huntington's Disease. Austin, TX: RG Landes Co., 1994.

59. Hantraye P, Riche D, Maziere M et al. Intrastriatal transplantation of cross-species fetal striatal cells reduces abnormal movements in a primate model of Huntington disease. Proc Natl Acad Sci USA 1992; 89:4187-4191.

60. Peschanski M, Cesaro P, Hantraye P. Rationale for intrastriatal grafting of striatal neuroblasts in patients with Huntington's disease. Neuroscience 1995; 68:273-285.

61. Shannon KM, Kordower JH. Neural transplantation for Huntington's disease: Experimental rationale and recommendations for clinical trials. Cell Transpl 1996; 5:339-352.

62. Šramka M, Rattaj M, Molina H et al. Stereotactic technique and pathophysiological mechanisms of neurotransplantation in Huntington's chorea. Stereotact Funct Neurosurg 1992; 58:79-83.

63. Kurth MC, Kopyov O, Jacques DB. Improvement in motor function after fetal transplantation in a patient with Huntington's disease. Neurology 1996; 46:A274.

64. Goetz CG, Bakay RAE, Fine A et al. American Society for Neural Transplantation Registry for fetal mesencephalic implants: Demographic and baseline data. Abstr Am Soc Neural Transpl 1996; 3:25.

65. Boer GJ. Ethical guidelines for the use of human embryonic or fetal tissue for experimental and clinical neurotransplantation and research. Network of European CNS Transplantation and Restoration (NECTAR). J Neurol 1994;242:1-13.

66. Wolfslast G. Legal aspects of neurotransplantation. Zbl Neurochir 1995; 56:210-214.

67. Brundin P, Nilsson OG, Strecker RE et al. Behavioural effects of human fetal dopamine neurons grafted in a rat model of Parkinson's disease. Exp Brain Res 1986; 65:235-240.

68. Clarke DJ, Brundin P, Strecker RE et al. Human fetal dopamine neurons grafted in a rat model of Parkinson's disease: Ultrastructural evidence for synapse formation using tyrosine hydroxylase immunocytochemistry. Exp Brain Res 1988; 73:115-126.

69. Pundt LL, Jörn EA, Conrad JA et al. Phenotypic expression of

human fetal cerebellar cells following transplantation into nude mouse cerebellum. Soc Neurosci Abstr 1996; Vol 22.

70. Jones WHS, ed. Hippocrates (460-375 BC), Vol 1: Epidemics A:XI. London: William Heinemann Ltd., 1972:164.

71. Björklund A. Intracerebral transplantation: Prospects for neuronal replacement in neurodegenerative diseases. Res Publ Assoc Res Nerv Ment Dis 1993; 71:361-374.

72. Dunnett SB, Björklund A, eds. Functional Neural Transplantation. New York: Raven Press, 1994.

73. Sotelo C, Alvarado-Mallart RM. Cerebellar transplants: Immuno-cytochemical study of the specificity of Purkinje cell inputs and outputs. In: Björklund A, Stenevi U, eds. Neural Grafting in the Mammalian CNS. Amsterdam: Elsevier, 1985:205-215.

74. Sotelo C, Alvarado-Mallart RM. Growth and differentiation of cerebellar suspensions transplanted into the adult cerebellum of mice with heredodegenerative ataxia. Proc Natl Acad Sci USA 1986; 83:1135-1139.

75. Björklund A, Lindvall O, Isacson O et al. Mechanisms of action of intracerebral neural implants: Studies on nigral and striatal grafts to the lesioned striatum. Trends Neurosci 1987; 10:509-516.

76. Dunnett SB. Specificity of cerebellar grafts. Nature (Lond) 1987; 327:366-367.

77. Wu C-Y, Bao X-F, Zhang C et al. Fetal tissue grafts for cerebellar atrophy. Chin Med J (Beijing) 1991; 104:198-203.

78. Rosenfeld JV. Current issues in neural transplantation. Ann Acad Med (Singapore) 1993; 22 [Suppl 3]:464-469.

INDEX

Page numbers in italics denote figures (f) or tables (t).

[3H]flunitrazepam, 15
[3H]flunitrazepam binding, 54
[3H]GABA binding, 56
[3H]thymidine autoradiography, 101
28 kDa Ca2+-binding protein (CaBP, 101
28 kDa Ca2+-binding protein (CaBP), 131
6-hydroxydopamine, 89

A

3-acetylpyridine, 106
activity monitoring, 141
adenovirus, 86
alcoholic cerebellar atrophy, 38
American Society for Neural
    Transplantation, 157
Ammon's horn, 57
amyloid β-precursor protein (βAPP), 60
anophthalmia, 83, 84
anti-Leu-4 (CD3) antibody, 101, 106
anti-spot 35 antibody, 10, 101
aprotinin, 64
Arnold-Chiari malformation, 37
astrocytes/astroglia, 103-104
ataxia telangiectasia, 36-37
ATP-dependent glutamate uptake system, 61

B

basic fibroblast growth factor (bFGF), 16
basket cells, 5, 8, 13, 96, 139
Bergmann glia, 103-104
Biemond's hereditary posterior column
    ataxia , 34
blood-brain barrier, 100
Boder-Sedgwick syndrome. See ataxia
    telangiectasia
brain-derived neurotrophic factor (BDNF),
    16, 102
5'-bromodeoxyuridine (BrdU), 101

C

CA1 pyramidal neurons, 57
Ca2+-dependent cGMP, 17
CaBP, 9, 120, 121, 122, 123, 131
calcicludine high-affinity binding sites, 61
Capacitance field, 142–143
capacitance field, 143
cartwheel neurons, 48
cerebellar anlage, 10, 97, 114, 125, 151
Cerebellar ataxia with retinal degeneration,
    32
cerebellar atrophy of Holmes, 35
cerebellar degeneration of Marie-Foix-
    Alajouanine, 35
cerebellar degeneration of Norman type, 36
cerebellar glomeruli, 99
Cerebellar Graft, 96, 100, 113, 115, 117,
    119, 121, 123
Cerebellar graft, 105, 126
cerebellar graft, 1, 2, 89, 98, 99, 100, 101,
    102, 103, 104, 105, 106, 116, 118,
    119, 120, 121, 122, 123, 125, 132,
    133, 134, 136, 150
cerebello-olivary degeneration, 35
cerebellomedullary cistern, 115, 117, 118
Cerebellum
    higher brain functions, 144
cerebellum
    fetal, 102
cerebrospinal fluid, 100, 115
cGMP phosphodiesterase, 86
cGMP-dependent protein kinase, 56, 123
chimeras
    pcd, 50
    reeler, 57
    staggerer, 60
    wv/+, 64
    wv/wv, 64
β2 chimerin, 61
ciliary neurotrophic factor (CNF), 16
Clarke's nucleus, 106
climbing fiber, 8, 13, 46, 54, 59, 62, 63, 97,
    99, 102, 106, 119, 124, 138
CNQX, 15, 102

CNQX (6-cyano-7-nitro-quinoxaline-2,3-
   dione), 55
Cockayne's syndrome, 37
Congenital Malformations, 37
cortical histogenesis, 45, 58
corticospinal tract, 34, 35
Creutzfeldt-Jakob disease, 37

D

Dandy-Walker syndrome, 37
deep cerebellar nuclei, 5, 7, 8, 10, 13, 53,
   54, 55, 98, 117, 119, 120, 123, 126,
   128, 131, 132, 134, 138, 139, 141, 144
dehydrogenase of succinic semialdehyde,
   49
Deiters' nucleus, 53
delayed cavity transplantation protocol, 99
dendritic spine, 58, 62, 88, 105, 115
denervation-induced receptor
   supersensitivity, 150
denervation-induced supersensitivity, 139
dentate nucleus, 5, 34
Dentatorubral Atrophy, 35
Dentatorubropallidoluysian atrophy, 32
Diseases with Defective DNA Repair, 37
dopamine, 88, 152
dopamine (DA), 60
dorsal cochlear nucleus, 48, 103, 123
dorsal raphe nuclei, 8
Dysarthria, 36
dysarthria, 31
dysequilibrium syndrome, 37
dystrophin, 47

E

EGL, 61
emboliform, 5
embryos
   chelonian, 58
Encephalopathies
   Paraneoplastic, 38
encephalopathies
   paraneoplastic, 150
epidermal growth factor (EGF), 16
Equilibrium, 132
equilibrium, 15, 132
Ergometric activity, 142
ergometric activity, 142
external germinal layer (EGL), 7, 60
eyeblink conditioning, 50

F

facial nerve nucleus, 55
fascia dentata, 57
fasciculus gracilis, 34
fastigial, 5
Fink-Heimer method, 96
follicle-stimulating hormone (FSH), 86
frataxin, 34
Friedreich's Ataxia, 33
Friedreich's ataxia, 33, 34
frontal cortex, 63, 87

G

g-aminobutyric acid (GABA), 8
g-aminobutyric acid (GABA), 151
GABA, 14
GABAA receptor, 14, 15, 54, 55
GABAA receptors, 139
GABAA/benzodiazepine receptor complex,
   54
β-galactosidase, 104, 105
ganglioside GD1a, 53
GAP-43, 61
GCAP-8, 56
GDNF, 18
genetic engineering, 82
Gerstmann-Sträussler-Scheinker disease, 37
GFAP, 103, 104
Gillespie syndrome, 37
GIRK1, 64
GIRK2, 64
Girk2, 64
Glial cell line-derived neurotrophic factor
   (GDNF), 18
glial fibrillary acidic protein (GFAP), 103
gliogenesis, 140
globose, 5
GluR2/3 immunoreactivity, 102, 138
Glutamate, 15
glutamate, 14, 15, 18, 50, 59, 61, 102, 138
glutamate receptor, 15
   ionotropic, 15, 17
   L-AP4, 15
   metabotropic, 59
glutamate receptors, 102
   AMPA, 15, 102
   non-NMDA, 15
glycine, 15, 139
GnRH, 86

Golgi cells, 5, 8, 13, 96
Golgi epithelial cells, 104
gonadotrophin-releasing hormone (GnRH), 86
Grafts
   cerebellar, 96
   hippocampal, 86
grafts
   cerebellar, 98, 99
   ventral mesencephalic, 88
Granule Cell Layer Atrophy, 36
granule cells, 33, 36, 46, 47, 49, 55, 56, 57, 58, 59, 61, 62, 64
grip strength, 138, 142
growth hormone (GH) receptor binding protein, 16
guanosine 3¢,5¢-phosphate-dependent protein kinase, 101

H

habituation exploration behavior, 144

Hereditary spastic paraplegia, 34
hilus, 57
hippocampus, 86, 87, 98, 105
HLA histocompatibility complex locus, 36
horseradish peroxidase, 97
Horseradish peroxidase (HRP), 100
HRP, 100
Hsa, 33, 35, 36, 37, 64
human chromosome (Hsa), 58
Huntington's disease, 152

I

IGF-I, 16, 17, 48, 53
IGFBP2, 16
IGFBP5, 16
IGFR-I, 16
Immunopathological paraneoplastic degeneration, 38
Inferior Olivary Complex, 13
inferior olivary complex, 8, 13, 14, 35, 36, 46, 47, 53, 57, 59, 106, 117
inflammatory cytokines, 59
inositol phospholipid metabolism, 15
insulin-like growth factor (IGF)-I, 102
insulin-like growth factor-I (IGF-I), 16
interleukin 1b, 59
interleukin 6, 60
internal granule cell layer (IGL), 7

J

Joubert syndrome, 37

K

K+ channels, 61
kainic acid, 15, 101, 106
knockout, 65
Krabbe's disease, 37
Krox-20/lacZ14 transgenic mouse, 104

L

L-AP4, 15
L7, 56
laminin, 64
lateral septal nucleus, 63

N

nucleus of Deiters, 8

O

Open-Field, 133
open-field, 88, 136, 137
optic cup, 84
optic nerve, 84
oxytocin, 105

P

P-path monoclonal antibodies, 56
P400, 49
parallel fibers, 15, 46, 49, 54, 58, 59, 61, 62, 102, 115, 119, 123, 138
Parenchymal Cerebellar Degeneration, 35–36
Parkinson's disease, 88, 150, 151
parvalbumin, 139
Pax-2,, 18
Pax-3, 18
Pax-6, 18
PEP-19, 53, 56, 101, 127
Phaseolus vulgaris, 106
phenylhydantoin intoxication, 38
"pinceau" formations, 46, 124
plasminogen activator, 64
platelet-derived growth factor b-chain (PDGF), 16
platform accelerometer, 142
pleiotrophin, 16
plexiform layer, 57

'point-to-point' systems, 152
polypeptide, termed SP47, 59
polyribosomes, 50
pp60c-src(+), 59, 61
Primary atrophy of the dentate system, 35
Prion Diseases, 37
Purkinje cell, 122
    migration, 104, 119, 122
Purkinje cells
    migration, 10
Purkinje plate, 10, 11, 12
putamen, 35, 63, 149
pyramidal neurons, corticospinal, 34

R

ras, 99
receptors for w-conotoxin GVIA, 61
red nucleus, 8, 35
reelin, 58
retina
    pigment epithelium, 85
    transplants, 84
retrorubral nucleus, 63, 87
retrovirus-mediated oncogene transfer, 99
RORa, 60
rota-rod tests, 133
rotational asymmetry, 88, 141

S

S-antigen, 85
saxitoxin-sensitive Na+ channels, 61
Serotonin, 105
serotonin, 119
Shy-Drager syndrome, 35
spatial navigation, 50, 144
spatial working memory, 144
Spinal Degenerations, 33–34, 34
spinocerebellar ataxia 1, 35
spinocerebellar ataxia type 5, 58
Spinopontine degeneration of Boller-
    Segarra, 34
stellate cells, 13, 57
striatonigral, 35, 63
striatum, 87, 88, 89, 100, 104, 106, 149
sub-Purkinje plate, 10
substantia nigra, pars reticulata, 63

sulfonylurea receptors, 61
superior colliculus, 84, 85
synapse formation, 58, 114, 116
synapsin, 101, 116
Synaptogenesis, 96–100
synaptogenesis, 8, 13, 16, 49, 58

T

T lymphocytes, 55
TAG-1, 59
tenascin, 104
thalamus, 8, 35
Thy-1.2 antibodies, 106
thymus, 60
tonic/clonic convulsions, 63
transgenic organisms, 82
trinucleotide repeat expansion, 35
trk proto-oncogene, 18
trkA, 18
trkB, 18
TyrOHase, 63
tyrosine kinase receptor family, 18

U

uvula vermis, 115

V

v-myc, 99
ventral tegmental area, 63, 87
vermis, 48, 62, 122
vestibular nucleus, 8, 53
vimentin, 103

W

Wnt-3, 61

X

X25, 33, 34
Xenopus oocytes, 64
xeroderma pigmentosum, 37

Z

zebrin I, 101
Zebrin II, 56